WHO'S RUNNING YOUR FARM NEXT?

5 STEPS TO DEVELOP AND COACH YOUR NEXT GENERATION

WHO'S RUNNING YOUR FARM NEXT?

5 STEPS TO DEVELOP AND COACH YOUR NEXT GENERATION

SARAH BETH AUBREY

Niche Pressworks

Who's Running YOUR Farm (Next)?
ISBN: 978-1-946533-59-3 Paperback
ISBN: 978-1-946533-60-9 eBook

Published by Niche Pressworks; http://NichePressworks.com

Disclaimer

The stories and anecdotes in this book are my own, though based on many of the real clients, friends, and peers I've worked with or spoken to about these topics over the years. As such, names, specific events, locations, and other aspects have been invented or altered to protect the privacy of those involved.

ADVANCE PRAISE for
Who's Running Your Farm (Next)?

Other Works
by Sarah Beth Aubrey:

Non Fiction/Business

Find Grant Funding Now! (Wiley, 2014)

The Profitable Hobby Farm (Wiley, Howell Bookhouse, 2008)

Starting & Running Your Own Small Farm Business (Storey, 2006)

Fifty Ways to ACT Like a Pro (e-book, 2017)

Your Strategic Plan on a Page in One Hour or Less
(workbook, rev. 2018)

Fiction Series

Championship Drive (Bookbaby, 2016)

Heart of a Champion (Bookbaby, 2017)

Dedication

To the three most important farmers in my life:

For my Grandpa, Ray Willard, I miss you, but I know you're farming a beautiful piece of flat, black ground just over the horizon.

For my Dad, Charlie Potter, I will always see you as a farmer, and I love that.

And, for Cary Aubrey, my husband and partner, thank you for walking the pastures of life with me. I love you.

CONTENTS

A LETTER

Dear Farmer,

If you could write a letter to yourself back at the time when you first started farming, what would you tell that young person?

Would you tell them to stick with it and not to give up? Would you urge them to be more patient and work harder? Would you suggest they make bigger moves and take a few more risks? Would you tell them to ask the questions they put off and to have conversations early with those special old-timers who are now gone?

How would you have tried to prepare them for the times you've now seen? Actually, what would you have asked your predecessors?

You know you can't really go back, and you don't get the knowledge you now have for free. You've had to experience both the good years and the bad. But, do you realize how valuable your experiences are and how you can influence the future now? Now, you can answer a lot of the questions that you had for someone else.

Your footprints are already on the soil out there. This book is your chance to write your name on the future and use what you know and what you've accomplished to give your farm its best chance at success. Thank you for all you do for agriculture.

—Sarah Beth Aubrey

Part 1

FUNDAMENTALS

THE BIG SIX:
FUNDAMENTAL SHIFTS AFFECTING THE FUTURE OF AGRICULTURE

In less than five years, your farm will be very different than it is today. You probably won't be the one in charge.

Will Your Operation Make It to 2050?

You've probably got more questions about the future of your business than answers. Let's start with a heavy hitter: Do you expect your farm to be here in another generation, for example, in 2050?

Who will lead it there and beyond? Have you identified the next CEO of your business, and if so, is he or she preparing to operate a business that looks totally different than the one you run now, one that would be unrecognizable to your grandfather?

Research tells us that a massive number of farms will change hands during the next five years or so, meaning the future for agriculture is literally *now*. During the next handful of years, the industry will experience a mass exodus of key players from farming—some operations will experience succession, and other operations will fold. Is your operation one of the nearly 60 percent of farms that will see a leadership change of some kind? Agribusiness, which includes our input suppliers and production partners, mirrors us on a fundamental level—their leadership is going through a very similar shift.

Most farmers today say they want to see their operations continue, as many as 80 percent of those surveyed want their business to outlive them, yet many growers facing retirement tell me that they don't believe that their next generation is prepared to take over, and that's *if* they have younger producers willing to farm for another 20 to 40 years.

While savvy farmers have already implemented sound, measured succession plans outlining details of ownership, partnership, and estate planning—certainly essentials for the future—these tools don't go far enough. These strategies ensure a proper business structure is in place yet don't replace the human elements of business sustainability.

Failure to prepare the next generation to manage a business that will look very different from yours is like not feeding livestock—they'll get behind and be unable to catch up and could even die. Even if you've created a very nice succession plan, if you do not intentionally cultivate excellent leaders for the future, your operation risks extinction.

> **Even if you've created a very nice succession plan, if you do not intentionally cultivate excellent leaders for the future, your operation risks extinction.**

The Future – "If We Don't Change, You Won't Have One."

Your knowledge can help prevent this fate. But avoiding the strategic hard work, it takes to grow leaders that can handle 21st-century agriculture is a direct route to extinction. Developing, training, and coaching these leaders won't be automatic. It will take a long view and a concentrated effort that will include hard decisions about the role employees, *including family*, play in the business.

It's reportedly been said that the late Southwest Airlines CEO, Chair, and co-founder, Herb Kelleher, was once quoted as saying, "What we're talking about here is your future. If we don't change, you won't have one." That sums it up for agriculture as well. We have always adapted to changing technology and production practices. Where we lag behind other industries today is in our human capital development.

Fundamental Shifts in Agriculture
Six Reasons to Adapt

The Big Six Fundamentals below aren't just statistics. They are hard-hitting truths, tectonic plates shifting, much like how fault line-riding California is supposedly about to detach and float out to sea. These shifts may affect you immediately, or it may be half a generation before the true implications of these changes are seen in your business. But these factors are unavoidable. Your operation's sustainability depends on your understanding of these shifts and your preparedness to meet the changes.

However, there is some exciting news here. For each one of these shifts, opportunity exists. It's a silver lining, if you will, or the chance to capitalize on these inevitable changes in a way that can benefit your business. The Big Six could be considered threats or opportunities, but that's up to you.

One: Massive Transition in Current Leadership and Land

A multitude of significant factors fall under this leadership and land shift, not the least of which we've heard about *ad nauseam*—the average age of U.S. farmers. To recap what is commonly known, USDA reports in the latest (2017) census of agriculture, that the "average age of all producers is 57.5, up 1.2 years from 2012."[1]

The age of those farmers identifying themselves as "principal operators" trends even higher, at an average age of 65.[2] Some may believe concerns about age relate to a desire to retire or declining agility, making it more difficult to perform physical tasks. These are relevant, but the real issue is knowledge transfer. Or, more concerning, the absolute loss of knowledge if we fail to transfer it effectively.

> The real issue is knowledge transfer. Or, more concerning, what happens to the knowledge if we fail to transfer it effectively.

Stereotypically, younger growers may have a better handle on critical operational aspects, such as technology adoption. But local knowledge, long-term relationships (see number 6), and loyalty to suppliers basically ends when this aging farmer group passes on or fully retires. Another critical aspect of the aging agriculture population is land management. Older growers own or manage a huge portion of the farm ground in production. How quickly that land transitions can be difficult to predict, but the probability that it will occur "soon" is assumed by experts.

Agripulse.com reported that during the 10–20 year period ending in the 2030s, at least 370 million acres of agricultural lands will change hands.[3] This number matches up closely with USDA's estimates that farmers older than 65 control a lot of ground today—about 320 million acres, more than one-third of U.S. agricultural land.[4]

If you're considering retirement, you'll have plenty of company. The USDA further estimates that between now and around 2030, 500,000 farmers will retire or otherwise be out of the business.[5] That number is staggering given another "fact" that is so commonly cited in the industry that it has become an anecdotal "fact" that we've all heard before: people employed in agricultural work comprise less than 2 percent of the U.S. population.

Silver Lining Opportunity

Yes, the impact of so many ag producers and industry leaders leaving the industry will be widespread, but it will be felt primarily one operation at a time. The silver lining opportunity of this shift, however, is that the next generation can have their turn to come into leadership and run the business the way they believe it will perform best. To succeed, they will need the guidance of their forebears. For many, there is still time to share knowledge, contacts, relationships, and the little anecdotes that helped them be successful along the way. Take time now to ensure the transfer of this valuable knowledge.

Two: The Changing Face of Labor

There are several major shifts regarding labor. First, competition for labor is intense. A 2018 Agcareers.com survey of 100 agribusiness managers found that a whopping 70 percent reported that their number one human resource (HR) issue involved competing for the same workers, including the challenge of setting themselves apart in recruitment.[6]

It is not easy to entice someone into farm work, especially today when most people are far removed from production agriculture. The appeal of a labor-intensive job with high seasonal hours is diminished when compared to a job with regular hours, a more comprehensive benefits package, and a climate-controlled work environment.

Second, understanding how to recruit and hire workers from outside the U.S. is becoming a necessity. While fairly common in dairy and vegetable crop operations, farmers in certain sectors (such as row crop) have not traditionally pursued these kinds of workers. Learning to navigate the regulations—and the language barriers—is a new experience for many.

Third, the labor pool may be a lot more feminine than you think. In developing countries, women *are* the ag labor force; the Food and Agriculture Organization reports that it's not uncommon outside the U.S. for women to comprise 40 to nearly 70 percent of farm labor.[7] Have you considered maternity leave for your tractor drivers, harvesters, and milking technicians? If not, it may not be long before you do. As international labor becomes more common, your next generation may even hire more women than men to do general farm tasks.

> **The face of labor is changing; it may look a lot more feminine than you think.**

Besides your general labor pool, your leadership pipeline will likely look decidedly more feminine as well. This is already reflected in hiring practices across the industry. In fact, fifty-two percent of respondents to an Agcareers.com survey indicated that they had hired more women than last time they were surveyed.[8] This change is not likely the result of a mandate or even a preference to hire more females. Rather, this shift could be because more young women are qualified and available to hire than ever before.

Today, more women graduate with college degrees than men. Women also earn more *agriculture* degrees in majors that are both science- and business-oriented. In an article from *The Wisconsin*

Farmer citing statistics from the National Center for Educational Statistics, we learn that, in 1980, women earned about 30 percent of agriculture degrees. Just 40 years later, women make up *more than half of the graduates.*[9] Purdue University's College of Agriculture reported that its 2015 enrollment included a split of 55 percent women to 45 percent men with the trend continuing to grow toward more women enrolling in the college than men.[10]

> **Going forward, personnel managers of the general farm labor pool must be trainers and will need to possess foreign language skills.**

The cost and diversity of both quality executive talent and general labor are changing fast. Emerging CEOs must become experts at fostering a culture where farm operations are an appealing place to work. Going forward, personnel managers must be trainers and will benefit greatly from the ability to communicate effectively with workers and leaders from other countries. Have you considered that fluency in a second language should be a requirement for your next CEO?

Silver Lining Opportunity

Don't be afraid to hire women. That's the lesson for operations that haven't quite figured out where or how to fit females into leadership or have just never had the opportunity in the past. Successful operations benefit from a variety of viewpoints, education levels, backgrounds, expertise, and work styles.

Find the best potential fits for a position and give them every possible opportunity in a conscious, measured way. In agriculture, farming forward is going to require all the very best hands on deck, just realize that you may need to add a few pairs of size small gloves to the shop drawer. Dads, I hope you don't wonder if your daughter *should* have a place on the farm someday. Actually, your chair looks like a great fit.

Three: Global Marketplace – A Chickpea Story

Let me start with a question: When you opened your lunch pail at school, did your mother pack you hummus with jicama sticks? (Are you googling jicama right now?) Mine certainly didn't. I'm a solid, 1970s-born Gen Xer, and the funkiest thing I ever had in my *Little House on the Prairie* lunchbox was a half-eaten bologna sandwich left over from the weekend. Yet my 10-year-old niece has this treat regularly.

Actually, I rather like hummus and the pre-packaged, pre-cut jicama sticks I get at the local grocery. Now, so does my friend, Pete, a farmer from Iowa. Two years ago, he started growing chickpeas—also known as garbanzo beans, the base ingredient in hummus—on contract for a global producer of the savory spread. Pete has been a corn farmer for 30 years. But recently, he replaced much of his regular No. 2 yellow dent corn production with chickpeas and converted a dedicated space in his new grain facility to segregate and house the product. His chickpea buyer comes directly to him to pick it up. Once he penciled out the profits, Pete has never looked back.

It is also well-known that global crop production has been steadily increasing in recent decades. These competitive pressures on traditional row crops, such as corn and soybeans, are only increasing the glut of U.S. grain as South America continues to produce and Eastern Europe gains traction. As an industry, we know these competitors are not going away, but adapting our crop mix is a difficult sell. After all, just two generations ago, small, diverse farms gave way to larger, specialized operations—a trend that continues in developed nations. Yet, with the increase in buyers and premiums for nontraditional crops, your next CEO will have to take a hard look at diversity again to remain profitable.

Hops for craft beer and barley for distilling? My clients Jeff and Jeremy from New York are already doing a small amount with local companies.

Hemp for fiber and CBD oil? Maybe. My friends Deb and Jerry in Colorado are going for it and have created their own brand.

When you opened your lunch box at school, did your mother pack you hummus with jicama sticks?

How much could the farm switch up the crop or livestock mix year-to-year to turn a better profit? This shift in thought process is not easy, inexpensive, or at all convenient. Yet, owning a massive fleet of equipment and keeping drivers on during the off-season may give way to outsourcing things such as spraying and harvesting in order to remain flexible so you can adjust a portion of the operation each year as global conditions dictate. Innovative growers know that alternative crops that appeal to consumer's changing global food preferences can be a financial boon.

Silver Lining Opportunity

Leaders with a keen interest in international food trends and an ability to forecast effectively may be rewarded for their efforts by getting to know these markets and meeting the buyers to understand the potential for production. State and national trade organizations and the government have traditionally been the route to opening new markets, and they still are. Yet, savvy growers know they must get involved to understand how to adapt to trends and threats. Things like leadership programs, international travel, and serving on or supporting your state and national boards are all part of our global role. The great news is that younger generations often love to innovate because it is their chance to try something new or to demonstrate to their peers and elders their own unique abilities and talents.

So, have you named an international ambassador for your farm yet?

Four: Consumer Preferences, Perceptions, and Power

After a look at global markets, it only makes sense to point out the burgeoning effect of consumer preferences on our business. Have you suddenly found yourself sitting across the Thanksgiving table from someone who knows absolutely nothing about agriculture? Or, has your niece (or daughter) gone away to school only to return having decided that GMOs are the devil, meat consumption should be banned (unless created in a petri dish!), and almonds really do make "milk"?

Sound like a horror story? It's not. It's a reality of which we are all painfully aware. We could blame it on millennials (we blame this much-maligned generation for just about everything else). But instead, we should realize that *every generation* influences the kinds of food that are produced by creating more demand for some foods and less for others. Call it whatever you want, but don't call it unimportant, because with the power of the internet, the impact of changing food preferences is happening faster and faster and reaching more people than ever before.

Even if we don't consider ourselves in direct-to-consumer sales, consumer preferences matter. Our customers—be they processors, millers, or packers—give tremendous weight to the emerging views of food consumers in the U.S. and around the world. We can be angry about this—in agriculture, we've spent a fair amount of time complaining that the rest of the population doesn't "get what we do for them." But, again, it doesn't really matter; we have to accept change if we hope to be in business in 2050.

Silver Lining Opportunity

Looking forward, what's the job that's needed here? Is it an ag advocacy role that helps debunk ridiculous myths or educate consumers? Or does it just require strategically evaluating what we raise to take advantage of higher profits if something different than what we currently raise is what people would rather buy?

That's for you (and your next generation) to seriously figure out—it's necessary to be ahead of the curve instead of lagging behind and fighting it. Will you need a marketing person or a social media manager? Or, perhaps you name an advocacy and community outreach point person. Maybe you share an outsourced marketing agency with another operation or two (or ten)?

Your next CEO has the opportunity to influence and educate. While that might sound like a lot of work to you, younger generations are often comfortable and feel a sense of responsibility to build bridges with non-ag consumers. They have a clear, native understanding of the power of an internet-enabled device, not only to access information but to change the world.

Five: Changing Technology, Data, and Online Landscape

Everything written in this section is already out of date—at least it seems that way with regard to how fast information can be transferred today. We all know the impact of the internet-enabled world will be magnified immensely for the next generation of leaders. Though still an issue, broadband access is steadily improving in rural areas, as rural telecommunication providers and even rural electric cooperatives strive to make investments in fiber optics and expand the availability of gig internet service. In a 2018 report on industry trends, ag lender, CoBank, reported a "bullish" outlook on all things data-driven in rural areas. Specifically, they indicated, "Applications such as 5G, augmented/virtual reality, and edge computing will continue to fuel growth in data traffic".[11]

An uptick in the option to work remotely is already occurring, enabling farmers to use professionals who don't even work at the farm to aid in farming itself. And virtual reality may allow for business consulting and even equipment repairs (yes, seriously) via the internet. Farm leaders going forward will need a better understanding of the value of data generated from monitoring and tracking systems on the farm, and the sensitivity not to share this information too carelessly or too frequently with the multitude of vendors offering the latest free software or a gadget to collect it. You'll need to become more selective about whom you work with to protect your important data assets.

Technology use also becomes a personnel management issue—how is your current expertise with public relations and crisis communications? For most operations, these skills will need brushing up (or even may require hiring a PR specialist on retainer). Today, employee use of social media during work hours is not only a safety issue when operating equipment but also a serious concern when a thoughtless social post piques the interest of PETA or the local news. Going forward, farming will require managers who are trained and capable of building social media use policies and handling data breaches.

Silver Lining Opportunity

Good news, technology is cool. And degrees in technology—beginning with greater efforts to focus on STEM (science, technology, engineering, and math) starting in primary school—mean your future generations are not only interested in these areas, they are far better prepared to lead than you are.

This type of job and skill set are already a reality for many in agriculture as we have become accustomed to using technology to manage inputs. But this sector continues to grow for anyone planning to farm in 2050. Consider factors such as increased contracts with buyers and the increased attention to the need for traceability and segregation or crops and animals. So, in evaluating roles of the future, have you considered an on-farm data analytics specialist?

Six: Relationship Shifts – Landlords, Locals, and Lenders

As one generation retires and the next one takes the helm, one the of the biggest shifts in agriculture will be how effectively relationships adapt to this change. From landowners to landlords, local banks to local cooperatives, the landscape is evolving about where and who we do business with—or without.

Landlords: We've already addressed changes in land ownership coming as many growers exit the business. But this also relates to landowners, either those we currently rent from or those we will be renting from in the near future. If the average age of growers is over 58, many of you rent from landowners that average age 88! We know the landowner's heirs are two, or even three, generations removed from the land, and they may have very different views about agriculture than we do.

How much training has your next generation had with negotiation skills and contract development? Handshake deals between farmers and landowners no longer work when there is no longer a person down the road we can shake hands with. The time to begin transferring the relationship management of this critical aspect of your business is now. Do you and your successors know who will inherit the land you currently rent? Have you met with them? More importantly, has your son or daughter?

Locals: Many of us already face situations where we no longer know our neighbors. More than ever before, our farming decisions are at risk of being scrutinized by uninformed locals, and some of us also already face resistance from fellow farmers, especially if livestock expansion is a consideration.

In your operation today, who handles public relations? Are you currently the only face of the brand with neighbors, zoning boards, and community groups? If so, that's a risk that can and should be mitigated sooner rather than later. Sometimes, especially when we've been in a community a long time, there are old wounds. So, in addition to creating new relationships, your next CEO should be working her magic to build bridges and expand community support for your operation.

Local suppliers of inputs (such as crop protection and seed) and local buyers of livestock have traditionally played an important role by providing the needed partnership, access, convenience, and the security of a local market and needed inputs. But, while the quality these providers supply has remained constant, much of this model has eroded or at the very least been consolidated in the last generation.

At issue here is *your loyalty*, or should I say *the likely lack of loyalty* your next generation likely feels toward suppliers. But your suppliers know that the future of Granddaddy's co-op is uncertain, and they are well aware that younger, more tech-savvy buyers are more likely to go online—whether to price shop or to replace the products they can buy down the road.

The cooperative model has long been held as a system that pays members back with equity or patronage. But my own studies with cooperatives show that younger members who've never seen these benefits perceive very little value in them and have a limited understanding of co-op structure. If your parents are the ones still receiving the cash and equity, your next generation finds that simply intangible. I conducted a survey on this topic recently, with one 30-something grower quipping: "Value in equity? Yeah, right, that's 80 years away, and two people have to die first."

The perceived value of suppliers is lower—or at least different— for younger generations, and that is a situation the supply chain will have to sort out with you. Yet, as seasoned operators who understand the value of loyalty to local partners, it's still important to encourage the younger generations to get to know these suppliers and build relationships with them. In fact, these organizations may be places to look for different kinds of strategic relationships and even improved operational efficiencies with regard to equipment and employees.

Does your successor look at their suppliers as vendors only or as possible partners in the business? Considering all these factors, it's essential that you consciously work to foster new, emerging local relationships with local suppliers where you're likely to find a lot of willingness on their part. Building loyalty with the next generation of your farm has never been so important to them as it is now.

Lenders: The third shifting relationship is the one with your banker. Many innovative growers and ranchers today have already outgrown the size of their local banks and work with a financial representative who is states away and visits infrequently. Does that representative personally know your "next in line"? Has your young leader already established credit with them? And, does that relationship feel like a partnership?

If it hasn't happened already, it won't be long before your banker is your son or daughter's age, not yours. Once considered pillars of the community, today, local banks are becoming relatively rare. If you still have a quality local bank in your community, consider yourself fortunate—and take a good look at them. The likelihood of your banker being a neighbor is certainly a lot lower in most rural areas today, and it's going to become even less.

A 2017 *Wall Street Journal* analysis of federal data showed that out of 1,980 rural counties in America, 625 of those no longer have a locally-owned or community bank. This number has doubled since 1994.[12] Banks have moved or merged. And though not always the case, these new bank representatives often don't understand agriculture and rely less on your history and reputation and more on strict algorithms to make lending decisions.

It's unlikely that your next generation will have the same long-term relationship with key banking advisors that you have had, so now is the time to demonstrate strategy and a clear path of succession that the lenders I've spoken with are telling me is so critically important.

Quality financial data and a strong balance sheet are the means of obtaining capital for the next generation. Does the next person coming on board need more training in accounting than you had? Are you making time to prepare your next generation to be savvy financial managers? Even if you have strong banking relationships today, they are not at all assured as banks consolidate and "your person" moves on. Your successor needs a solid understanding of what financial institutions will require and how to prepare it.

Silver Lining Opportunity

These shifts open up new doors and the opportunity to form new strategic relationships that can help both growers and suppliers stay in business. As some growers exit the day-to-day management of farming, agricultural land is naturally up for grabs. Will that land be sold or developed? Either way, it's an opportunity for the well-positioned grower to expand via rent or ownership. The best way to capitalize on opportunities is to prepare younger generations to be involved in strategic relationships now.

Time for a Solution

While the Big Six Fundamentals will affect everyone in agriculture soon (if not already), the real importance of looking at these topics is first an awareness of the potential impact they will have on your operation and then preparedness. Thinking about these big shifts often leaves people feeling overwhelmed, even if the change presents an exciting opportunity, which many of these do.

What can you do? Take action instead of being stymied by indecision in a place of not being ready to do something yet waiting for something to occur. It's a tremendous risk when we are kicking the tires with thoughts like:

"Maybe one of the kids will come back."
"Maybe my son will quit his job and decide to farm."
"Maybe I'll just quit and sell this place."
"Maybe we'll figure it out when we have to."
"Maybe I just don't like to talk about my finances with the kids."
"Maybe I don't want to face the facts that I don't know what to do."

The Big Six aren't trends; they are factors in our industry that are happening, regardless of our input and, certainly, without our control.

So, how do we have some measure of control and influence? We prepare for the future by creating it intentionally.

This book isn't your typical succession planning book. Rather, it's a book about the decisions you need to make, and the training, coaching, and development investments needed to meet future challenges and to capitalize on opportunities. What you need to do is take the initiative and build capacity on your bench to create an operation, of any size, that is proactive, not reactive.

Who's Running Your Farm (Next)? isn't a wakeup call. It's a call to action. Succession planning is essential, and there are excellent professionals and resources available to put together a business structure that works for you. But with this book, I'm asking you to peel back the layers of the *people side* of the equation so that you can begin asking questions, making decisions, and implementing plans to grow the right people for the roles that the future will require them to play. I am asking you to take time to examine the knowledge, skills, and leadership abilities of your emerging leaders and then make time to get them to the next level—a level you may not even be able to picture but must realize is ahead.

> **Who's Running Your Farm (Next)? isn't a wakeup call. It's a call to action.**

This book is here to challenge and encourage you by providing a set of ideas and strategies that will set you on the right course to consciously uplevel the human capacity of your business. I want you to be sustainable to 2050 and beyond. So, let's begin now with a process that you can implement.

HOW TO USE THIS BOOK:
FIVE STEPS FORWARD

Leadership SOI (L-SOI)

You're infinitely familiar with the idea of ROI or return on investment. While there are plenty of resources with advice about how to improve the return on assets and inputs, this book focuses on leadership and builds the case for investing in leaders now as a critical part of your farm's sustainability plan. I call this process your Leadership SOI or L-SOI, and it's divided into three parts—Self Leadership, Operational Leadership, and Industry Leadership.

Self Leadership relates to self-governance and self-awareness.

Operational Leadership is about understanding how to run the business and how to place the right people in the right role.

Industry Leadership is about the obligation to lead by playing a part in innovation, creating dialogue with consumers, and cultivating sustainability.

The L-SOI concept is based on the well-known premise that great leaders are not born; they are made. If you believe that, as I do, then taking time to assess and develop your business' L-SOI is a step toward helping others emerge as the leaders for the next generation.

5 Steps to Develop and Coach Your Next Generation

Step One: Start with Why. When you get clear about Your Why, you convert some of the reactionary, tactical-only thinking to a process of uncovering (or reconnecting to) your purpose. Then, it's much easier to outline a vision, and ultimately, a strategy that includes a set of goals for this process. You'll leave this section using some of the aspects of my proven strategic plan-on-a-page process that will result in a mini plan you can use to move forward with for the balance of this book.

Step Two: Take Inventory. Use Step Two to create your human asset state-of-the-state and consider what will be needed. What does your bench look like? How can current employees become better prepared to lead? How much real time is being spent identifying critical topics for knowledge transfer and taking the actions to do it? We'll look at the roles you have now and uncover those that are coming—especially the roles where no one is occupying the post—yet.

Step Three: Cultivate Culture. The idea of coaching—beyond the sports field—is fairly unfamiliar to most people in agriculture. This has been true both at the farm and in agribusiness, but it can't be the case any longer. This third step will define executive coaching and, most importantly, establish steps for intentionally fostering a positive, productive culture for your farm and family business.

Step Four: Build Influence. This section covers a variety of leadership development options for the next generation. The overall theme is about building a powerhouse professional network. I will cover topics such as training and education, the value of outside influences like peer groups and advisory boards, and how to gain global experience, all with an eye toward developing advocacy and communications skills.

Step Five: Develop Accountability. No successful process is ever complete without implementation. Yet implementation doesn't happen without good leadership and accountability at all levels. It's a joint effort. I will cover how to build accountability systems (like performance metrics and personnel reviews) as well as how to build career paths and understand when key employees and family members may benefit from lateral moves to get the necessary experience.

Future-Focused Leadership

It's important to know what you'll receive in exchange for your time investment in this book. Succession planning and cultivating long-term sustainability does not happen through the efforts of one individual alone. The best way to benefit from the five-step model is through collaboration. This book is intended for leaders at both ends of their farming tenures and provides a framework that you can model and modify to your benefit. But, it's up to you to act on it. Start by asking yourself, "Who is leading the charge on coaching and training today?" Do I have a path to provide "CEO training"?

> **Start by asking yourself, "Who is leading the charge on coaching and training new leaders today?"**

Who's Running YOUR Farm (Next)? is the start of a movement to get us, headed where we need to be as an industry, specifically related to building capacity and capabilities in people. It is a compilation of twenty years of experience working directly with farm families, cooperative and agribusiness boards and teams, and association leadership. The five steps that have emerged out of my experience will enable you to develop future-focused leaders who understand the difference between strategy and tactics, ideas and goals, good intentions and actions—and know when each applies.

Both those who have weathered the tough seasons and young people with boundless ideas and enthusiasm must pull together to become a strong leadership base that allows the business to thrive and stay relevant for another generation.

Part 2

A WAY FORWARD

START WITH WHY?

What's Your Why?

Did you ever take a nasty spill off your bike when you were a little kid? Of course, you did. We all did. Our mothers would scoop us up and ask, "Where does it hurt?" The answer to this question still tells us a lot about a situation. But, unlike falling off a bike and scraping a knee, the pain points in business aren't always obvious.

In agriculture, we're conditioned to just "handle it," putting aside uncomfortable issues or long-standing problems in favor of tackling immediate needs like harvest season or calving. Yet, such reactive thinking is not the way of the future. Proactive, forward-looking approaches are necessary to meet new demands. The first step in building a strategy to coach people effectively is to understand Your Why.

Simply put, Your Why is your purpose and how you want to see that purpose fulfilled. For many farmers, farming is a life's work; it's who they are. It's their identity. Does that describe you?

If you want to see the fruits of your labor sustained to 2050 and beyond, well, *why?*

Have you asked yourself lately why you farm? Depending on the season, I am sure you have *questioned it* many times—but this is not at all a tongue-in-cheek or sarcastic question.

Why do you want your farm business to outlast you? Why does it *really matter?*

One of the most common "why" responses I hear from growers is explained in the retelling of an experience they had earlier in life. It's often found in the situations and problems they faced—and overcame—that their why emerges. Consider Bruce's story.

Have you asked yourself lately why you farm?

Why Bruce Farms Today

Bruce graduated from college in the early 1980s. He had a corporate agribusiness job and was doing the weekend-warrior-farming-thing so he could help out his dad and uncle. This was driving his young wife mad—when he wasn't working late, busting his tail for the company, he was always at the farm, two hours away from home. She complained that they never saw each other or did anything fun. Granted, they had three of their five children during those years, so apparently, there was some time for fun …

Anyway, being the 1980s, times were lean, and credit was extended. Dad and Uncle weren't getting along. And when Grandpa died, it got worse. Tempers were high; Bruce remembers seeing them come to blows one harvest over something as unimportant as who drove and who rode in the pickup.

As the credit crisis worsened, Uncle decided he wanted out. Dad honestly couldn't blame him, but there simply wasn't enough liquidity to divide it fairly between them. One day, Bruce found his dad dead on the tractor seat; he'd had a massive heart attack sometime between dinner and late afternoon.

Despite his grief over losing his dad, for a while, things actually got better. Bruce's uncle counted on him to stay and get the crop in. They even put one out together the next spring, though Bruce didn't get paid for this, his wife reminded him. Sadly though, at the end of that harvest, Bruce's sense of loyalty was "repaid" when Uncle decided to sell everything that fall. The land sold, the equipment sold, and Bruce wasn't prepared to buy back any of it. A full crop cycle went by without farming. But three days before their third child was born, Bruce told his wife that he'd left his job, and they were moving to the farm. He had put together enough of a grubstake to start small and buy it back.

(*Can you believe she stayed?* Of course, you can, because you either are married to a woman like her, or you know one. Or ten.)

Today, Bruce has been farming for thirty-some years. With tears of pride, he told me how he has literally reacquired every acre of the original farm and then some. He and his middle son, Jason, manage about 10,000 acres, and his son is setting himself up to be Bruce's successor. And, you can bet that they have a plan to ensure he's developing the skills needed! Bruce's wife, Janice, saw to that!

But Bruce agrees with the importance of creating a solid succession plan and clear development goals for Jason because he has always been very clear about his Why. You see, Bruce's goal has never been to leave his family land. Yes, that's correct. Instead, he aims to consciously co-create a sustainable business model that can be run by trained, competent people. Thankfully, this includes his son.

When he reached out to me because he had heard about executive coaching, Bruce already had a foundational plan that laid out terms of employment and established succession structures. But he also knew that formal structures and legal documents don't create and grow a 21st-century leader. It was a start, but it didn't go far enough. With his Why in mind, we worked through the steps you'll read about in this book. We started with a strategy where he and Jason could build goals, set milestones, and evaluate progress. Bruce has no intention of having his son experience the pain points that he did during transition.

What Are Your Pain Points?

For those of you like Bruce, starting with Your Why is very obvious. But for others, it can take some real soul searching and serious discussions with family. If Your Why isn't immediately obvious, don't get discouraged and skip this process. Instead, try looking through the lens of something of which you are probably well aware—your pain points.

Like acupuncture for decision-making, I use the questions in the sidebar during executive coaching conversations with business leaders and farmers alike, and these questions can really move the needle when a leader seeks clarity.

Points About Pain Points

Don't be alarmed, but I need to tell you that even the toughest among you may become emotional when you really begin to think about these questions. Maybe you already have. Don't let that stop you from doing the work. Hey, I always tell my farming peer group members that it's not a good meeting unless somebody cries. And, it's the tough guy every time—and that's okay.

If you are having trouble answering these questions, don't automatically think that you can't figure out Your Why. Consider asking the people closest to you to answer the questions and to weigh in on your behalf. We have mirrors all around us.

Finally, resist the urge to blow off this exercise because you are too busy to worry about it. This piece is foundational and worth it.

> ## *Where Does It Hurt? 5 Qs to Identify Your Pain Points*
>
> 1. Why do I farm (or why did I come back to the farm)?
> 2. Is farming the career of my choosing?
> 3. What is the most satisfying element of farming?
> 4. Why is farming important to me?
> 5. What happened during my transition into farming that I don't want to repeat for the next generation?
> 6. What is my greatest fear about the future?
> 7. What important business decision or action am I putting off?

What Are Your Results?

Suddenly feeling the need for a couple of Tylenol or a stiff drink (or both)? Very good. That was the point. View this effort like the no pain, no gain exercise slogan. Sometimes we get so good at tolerating our normal—even a normal that doesn't serve us now and certainly won't serve us going forward—that we push down the real pain and let it just simmer like a lifelong, dull ache. Using this book can help fix that burr in your side. If these pain points resonate with you, you may have found Your Why. Write it down.

If these pain points resonate with you, you may have found Your Why.

We are very programmed in agriculture to be task-oriented, to manage crises, and to meet uncontrollable deadlines effectively. Stopping to think about why we do something feels counterintuitive if you've spent a lifetime working with the mantra "because that is just what you do." But, therein, you might have found Your Why. Maybe it's that you value a strong work ethic or feel gratitude for all that has been given to your care. Or maybe you believe your time here is simply to be a caregiver of the land for the next generation. The Why often lies just below the surface, but it is still there.

Put down your smartphone and turn off that yield monitor. Step outside and overturn a few rocks. You'll find Your Why.

Put Your Money Where Your Why Is

Now, armed with a solid purpose, the entire concept and effort of creating an intentional culture of coaching has a lot more teeth to it. There is a reason—your reason—why doing something to prepare the next generation matters. Now is the time to begin crafting a strategy to get started.

The Long-Range Strategic Plan is Dead

I never thought I'd build a career around strategic planning, training, and executive coaching, but life has a way of providing opportunities where there is a need. I frequently hear things like:

"I'm not really a big picture thinker. I'm good at managing tasks."

"Do I really need a strategic plan? It's such a big cost of time and money."

"What's the difference between a strategic plan and a business plan?"

"I just don't have the time to 'do' strategic planning."

At the farm level, I've found that strategic planning has often been overlooked or considered something that "only big firms do." It has the reputation of being a very expensive, drawn-out, over-analyzed process, and rightfully so, I'm sorry to say. Yet, I am delighted to report that the old-school, long-range, ten-year strategic plan is dead.

There, I said it. And I'm a strategy consultant.

Beware the Binder Morgue

"Come with me," urged the VP's executive assistant as she pulled a set of keys from her desk. "It's time to introduce you to The Resource Room."

I followed dutifully; it was my first week at the consulting firm, and I was anxious to learn. Pausing, she approached a large metal door and then opened it almost reverently. I could feel the cold air brush past my face as we stepped into the darkness, and she searched for the lights. "Go ahead. Take a look," she urged. I stepped into the room. It stood silent as a stone against the noisy backdrop of the busy office culture outside the door. The room was nothing but wall-to-wall, 10-foot tall shelves lined with binders. All the big names in the industry were there, organized by alphabet and then by year. Strategic Planning Initiatives 1996, 1997, 1998, and 1999 for one major client took up nearly 20 feet of binder space. "That's one of the largest planning processes the firm has ever done," she commented proudly.

Dumbly, I asked: "So, what am I supposed to do with access to this room?" I hadn't meant it to come out rudely.

The executive assistant eyed me sharply as if I'd insulted the entire firm. "Well!" she huffed. "Like I already told you, this is the Resource Room! This is where client's binders go after the strategic plans are done." At lunch, I mentioned the Resource Room to a group of top-flight facilitators and trainers.

"Oh yeah, you mean the Binder Morgue?" laughed one of the guys aloud. Given the air of gravity the executive assistant had attempted to instill in me, I was surprised by this total lack of regard for "resources."

Another said, "Yeah, that's where all your hard work goes to die. Stay out of that place, and you'll be better off."

Over time, as the 1990s gave way to the 2000s and then the 2010s, the binders slowly (and thankfully) went away. First, they were replaced with cd-rooms and then came the flash drives. Now, it's cloud-based storage. But it's not just about reducing bulk. It's actually become less about crafting deliverables and more about focusing on accountability and implementation to get results.

The reality is that strategy is getting slimmer and more fleet. What it means to "do strategy" has evolved. Experts tell us there is no longer value in the ten-year strategy; the digital world simply changes too fast. It's becoming common in consulting circles to consider even five-year strategic plans to be dead. So, while leaders who focus on slow planning (or no planning at all) are comfortably sleeping, the marketplace is happily stealing the tires off their equipment, snapping up the best hires, and innovating as needed to meet demands.

Strategy at the Farm Level Now

So, if the long-range plan is dead, how do we "do" strategy now? What's the right time horizon? Some business books even tout the need for 90-day only plans. I consider these quarterly efforts excellent strategic check-ins but not sufficient for full-on future planning. In agriculture, especially as it relates to the topic of developing leaders, strategic cycles of 12–24 months, coupled with the 90-day check-ins, are an effective model. Some operations may opt to add a third year as a stretch time horizon, but that depends on the status of your leadership transition process.

It's important to note here that strategy and vision are not the same things, so don't confuse vision with setting up a plan to achieve your vision (the strategy). A vision for the future (and we'll get to that soon) is supremely important, but it's truly unreasonable and maybe impossible to create an exact plan for how you will achieve a far-off goal more than three years in the distance. Incremental steps, adjusted along the way for internal and external factors, are why the shorter strategic planning initiatives are now the norm.

Once you have Your Why, you're well on your way to understanding the very elements of strategic thinking. First, do you know if you're a solid strategic thinker? Do you consider yourself a details person and not a big picture thinker? If so, you're not at all alone. Many growers I work with in family board meetings confuse strategy and tactics. The place to start is with a definition.

What IS Strategy?

The original idea of strategy had nothing to do with business, binders, or executives. Rather, its roots are ancient and militarily founded. The great military minds collect intelligence and use that data to make decisions about how to deploy resources. It's easy to see how that definition translates to business. I rather like author Rich Horwath's definition, as found in the book *Elevate*. "Strategy can be fully defined as the intelligent allocation of limited resources through a unique system of activities to outperform the competition."[13]

Question: Would you describe strategic thinking as analytical or visionary thinking?

Got your answer?

What if I tell you that it is both? Surprised?

CliftonStrengths, a Gallup professional assessment product, helps leaders and teams best understand not only what their key strengths are today but how to leverage and maximize those for tomorrow. We'll talk more about assessments and how to use them in Step Four. Broadly, the "Strengths" concept categorizes 35 strengths into four areas—Executing, Influencing, Relationship, and Strategic. With regard to both analytical and strategic types, CliftonStrengths classifies them *both in the strategic category.*[14]

During facilitated training sessions, participants often fight me on this, saying things like:

"I'm strategic. I like to think about the big picture."
Or,
"She's analytical. She's a numbers guru."

These comments are only partly true. You see, strategic and analytical thinkers are both thinkers—people who process information before acting on it. People frequently mistype themselves or others as strategic when they are not. This happens in both agribusiness and on the farm. Actually, in agriculture overall, we trend toward being short on natural strategic thinkers!

Actually, in agriculture overall, we trend toward being short on natural strategic thinkers!

This opinion is entirely my own, stemming from observations after conducting thousands of assessments in the last decade. Whether on the farm or in the board room, I have found that a large majority of thinking in agriculture exists more in the executing realm and less in the strategic domain. Experience demonstrates that even the most highly successful operations of today are still more accustomed to *executing* what they already know how to do rather than pausing to consider and then implement a strategy that ensures the operation's future growth and success.

Whether on the farm or in the board room, I have found that a large majority of thinking in agriculture exists more in the executing realm and less in the strategic domain.

Individuals who execute tasks typically have a profound sense of responsibility. They are sometimes reactive (they are quick to lend a hand!) and task-oriented and are generally made up of natural born doers. Sound like anyone you know?

For much of history, this made sense for us in agriculture. "Make hay while the sun shines" was a viable adage to live by. However, staying too much in the task-oriented present, while we worry and ruminate about fears for the future, is dangerous to our business. Strategy helps us redirect that worry into something productive!

Build Strength in Strategic Thinking

Visionary, long-term thinking is a strategic quality, but so is the ability to analyze the steps in a process or the obstacles in a path, and then determine how those steps can be altered to meet a stated objective. Strategic thinking is like using a drone.

Drones capture big picture images (the full vision of the field), but they are of no value if the viewer can't interpret the data to make it more productive (the analysis). Knowing now that many people mistype themselves as strategic, does this mean that only people who have more strategic skills on the CliftonStrengths assessment should take the reins of the farms of the future? No, it doesn't. What is important is that all leaders learn to become more strategic in their thinking and make time to actively hone and then use the skill of strategic thinking.

> **Strategic thinking is like using a drone.**

Four Elements of Strategy

1. What is our organization's vision and the theory on which we operate?

2. What do we do? Or, what part of this market do we want to be in to deliver on what we do?

3. How do we do it? What actions, adaptations, and models do we create to be successful?

4. What capabilities do we need in order to implement what we do, such as skills of team, new or updated resources, and new or ongoing training for our people?

So, When Do You Want to Be Out of a Job?

Strategic thinking is more important than ever. So, make it a habit—a part of your culture. Where are you with your future people development strategy today? Have you set your exit goals?

Have You Set Your Exit Goals?

How do you want to leave farming?

Do you know who will be in charge of your operation in two years, five years, or ten years?

Are they prepared? If not, what conscious effort is being made to get them prepared?

Are they working in the business now? If not, is there a track to get them there?

If you're on the younger side of 40, how are you being groomed and educated to lead the farm for another generation?

What is being done to develop and coach the emerging leaders for your business?

Your Turn: Creating a Strategic Plan for Sustainability

Having done plenty of self-reflection now is a good time to re-involve other key people in the strategy discussion. In Step Two – Take Inventory, you will use the strategic direction established here to aid in your assessment of current and future personnel and training needs. During my years of leading small groups, boards of directors, and corporate work teams through change, I've developed a process called "Your Strategic Plan on a Page, In One Hour or Less." The plan-on-a-page concept is a strategic thinking tool for individuals or teams where one hour is set aside to work through each of five sections (more on that next) to create a strategic direction and a sense of agreement among everyone involved.

There are three primary reasons this process works. First, it's doable with little outside expertise. You don't need an outside consultant unless you want or need a moderator. I facilitate this exercise quite frequently, and there are certainly cases for hiring someone to coordinate the process. In fact, there are a variety of situations where it makes sense—when the family or work team is in major conflict or when you're looking to get started using strategic planning, and you want an outside resource to teach you, thereby building capacity in yourself and others so you can ultimately "take it from there."

Second, this process is a powerhouse for moving stubborn problems forward. The abbreviated timeframe requires your group to make quick, decisive moves. Each section is timed, and that is by design. Sometimes, it's not the problem that is stubborn. It's that the people involved in the problem's solution are stubborn.

Finally, the plan-on-a-page process combines a long-range view with short, actionable solutions and includes an opportunity to review at regular intervals. The process begins with the big picture and drills all the way down to selecting one priority with which to begin. I recommend short, frequent bursts of effective planning, followed by implementation and review. Some of the plan will change every time you schedule a review—often (and hopefully) those are the shorter term goals you have set. Other parts of the plan will not change, such as the core values for the operation.

> **Sometimes, it's not the problem that is stubborn. It's that the people involved in the problem's solution are stubborn.**

Strategic Planning Meeting Dos and Don'ts

Set the Stage

Many leaders opt to start the strategic planning process alone as a way to organize their thoughts before sharing with others, and that can be an excellent choice. But, when you're ready to invite others, prepare them. Properly positioning any new initiative is very important for maximum effectiveness to ensure that employees and others in the business don't feel overwhelmed with change.

Without proper planning and stage setting, you might find others on the employee team or family board are more inclined to schedule root canals than make time for your meeting. You need to establish a positive outlook because ultimately, successful strategy implementation has to be collaborative.

> **Ultimately, successful strategy implementation has to be collaborative.**

When selecting the key people to be involved, talk with individuals first, gathering perspectives and explaining Your Why one-on-one. Then, put together an agenda to distribute. Set the length of time, the arrival and departure expectations, and what everyone should consider advance.

Select an Appropriate Time

It's obvious that scheduling a strategy discussion should not be in the middle of harvest, planting, calving, or some other critical time. But there are many other factors that are going to distract people: holidays, vacations, summer break—the list never ends. And there you have one of the most common excuses for not taking time for strategy—no time.

There *is time*, but there is *no perfect time.*

Work with the people involved well in advance of when you want to schedule a strategy session and get the buy-in for the meeting. For anyone whose input is key, make attendance mandatory. You will also want to assign roles for this meeting. You could appoint a person to take notes, someone to be the timer, someone to make sure food is delivered conveniently, and someone to handle the follow-up actions (more on that later).

Set a Different Location

Yes, Grandma's kitchen table is convenient to the shop and the farm office. Yes, her cookies are delicious. But it is the wrong location for this meeting! The strategy discussion needs to feel like something different. This may be the most important tip of all: Go off-site! You could book a conference room at a hotel, ask your CPA or banker about using their meeting site, borrow the co-op's board room, or reserve a room at the library. Nothing diffuses a situation better than a change of venue. Bottom line: find a way to be away from the pressures of the operation and the distractions.

Also, make sure the key attendees are as focused as possible. Line up care for those with young children and manage employee's schedules to cover any workload responsibilities. And, for heaven's sake, have some decent food and beverages on hand! Hungry people do not strategize well.

Room set-up should be considered, too. The most effective set-up is boardroom style where everyone is around a large table and can freely see everyone else. Or, tables could be placed together in a U-shape to obtain similar results. Classroom style or separate round tables is not a good design for discussion and should be avoided. Remember, this is not a lecture. This is a group discussion.

The Plan-on-a-Page Instructions:

1. Before starting the hour, first divide up the five sections, assigning a time for each. Sample time idea (minutes): 5, 10, 20, 20, 5

2. Get your entire team assembled and ready supplies, like a whiteboard or pens and paper.

3. Use a timer or the stopwatch app on your phone and stick with it!

4. If you're doing this as a group, complete the project by having a debriefing discussion and compare notes. But, be careful not to let that derail the quick gut decisions you made during the hour!

5. Have someone summarize the messy notes into a nice plan-on-a-page that you can read, share, and use.

Want a template? Get my strategic plan-on-a-page worksheet or learn more about facilitated strategy sessions at SarahBethAubrey.com

The Five Elements of the Strategic Plan on a Page

I. Values: First, define the core values for your life and farm. What motivates you? Why farm instead of finding "a job"? What is this operation's legacy about? This question largely relates back to Your Why, but it asks everyone in the room to weigh in. While much of the plan will change over time, this section is often a baseline that endures through the ages. These are the God-Country-Family kind of values that people live by. Many operations already have this value statement hanging somewhere in the office.

II. Vision: Define your vision going forward. What are your aspirational dreams for your operation? Where is it headed under your tutelage? Take a look at the business today. Where do you see it emerging as the future unfolds? Do you see the operation expanding in one area and getting out of another? This is not a mission statement, though you may choose to write one from the ideas expressed in this visioning and values process.

III. Analysis: In this section, I commonly use the SWOT analysis system, going through each very quickly—the Strengths, Weaknesses, Opportunities, and Threats (SWOT). Give this section and number four the most time. One way to break this up effectively is to list only one to three items for each of the four components of the SWOT system. Do not skip any.

Strengths and weaknesses should be about things that already exist; they are internal and specific to your business, not market conditions. For example, strengths might be your land base or the health of your balance sheet. Weaknesses might be a lack of well-trained employees for certain functions or the distance between your locations. These may be things that can be changed by you, or they may be very difficult or impossible to change. In any case, they are unique to your farm.

As for opportunities, these are external. But they are things that you can affect and possibly take advantage of for your benefit. New family members coming back create an opportunity. The possibility of a new marketing terminal in your area is an opportunity.

Finally, threats are always external. They are very difficult (maybe even impossible) for you to change personally, and they do affect your business. Threats could be global economic crises, political changes at the federal level, or disconcerting regulations.

IV. Goals: Set both short- and long-term goals and give them some meat, using the tried-and-true, SMART method for each—Specific, Measurable, Ambitious, Realistic, and Timelines. While you may be tempted to create long lists of goals, don't get stuck in the weeds here. You can build out longer goals during other sessions or with individual departments or individual employees, as they relate to their development. Now is not the time for this.

For this exercise, you should develop two to four SMART goals for the operation, NOT a list of tasks to complete. For example, you might set a goal of improving the company's financial position. Then later, you can create the tasks that help you to achieve this goal, such as meeting with two new lenders or reviewing your budgets with a professional for areas of inefficiency. Because the plan is set to be reviewed in 90-days, you should strive to make real progress on some of your goals during that time, even if you are considering a strategy that overall may be your direction for 12–24 months.

> **There are many tasks to be completed and many goals to be achieved. But you can only have one priority at any given moment.**

V. Priority: Now, set the top priority. Just one! Priority and goal are not synonyms, so consider what Greg McKeown wrote in his book, *Essentialism: The Disciplined Pursuit of Less.* "A non-Essentialist thinks almost everything is essential. An Essentialist thinks almost everything is non-essential."[15]

Reread that quote to really let it sink in—*almost everything is nonessential.* There may be many tasks to complete and many goals to achieve. But truly, a person can physically only have one priority at any given moment. Work hard to gain agreement around the very top priority and then decide what time, people, and financial resources need to go toward this priority.

A Final Step: Accountability

Now that you've created the plan, use it! Accountability is the key to keeping even the most driven people on track. You can improve accountability by scheduling the first review before leaving your current strategy session and naming an accountability lead or quarterback to keep the plan alive until the next meeting (Hint: the best person for this job may NOT be you).

Follow up after the meeting with a request for feedback from everyone involved. Either do this via email or stop by to see your team in person. Then, share the major components of the strategy with everyone in the operation and help them to understand where the business is headed. Finally, you might also choose to assign accountability partners, as needed, for the tasks and goals that came out of the session. We'll cover more on this in Step Five.

Your Decision, Your Buy-In

Ever done something half-heartedly? How did it work out? How did you feel about yourself? Did others notice your lack of effort, too? Remember the Leadership SOI that I mentioned at the beginning of the book? This section is all about the first part of L-SOI—Self Leadership, which includes self-governance and self-awareness. Leading with awareness and thought requires not only strategy but also a clear vision forward.

Until you understand the importance of setting a vision and plan, you'll keep putting it off as other important, reactionary issues arise. But a clear vision is foundational, and it's sustaining. Don't leave this section without confirming *your own* buy-in. Developing the next generation is an ongoing process requiring a commitment to excellence. Your commitment needs to come *first* and set an example for others.

Strategic planning is no longer a huge process, done in painful bursts. It's also no longer an optional process saved for later. Strategy is never done; you cannot check it off your list and put it on the shelf. It's an evolving, important process that helps you to adapt as your world shifts. The binder room has been sealed off and archived for good.

TAKE INVENTORY

If you cannot risk, you cannot grow. If you cannot grow, you cannot become your best. If you cannot become your best, you cannot be happy. If you cannot be happy, what else matters?

—Dr. David Viscott, from *Risking*

When Biscuits Fly!

Iowa brothers Tom and Bob stood in the warmth of Grandma's garage enjoying a couple of beers and watching the snow fly across the prairie. All was (mostly) right with their worlds. They farmed together, and though it wasn't perfect, after five years, they got along pretty well.

They both had wives and little kids. The afternoon had been spent watching football and talking about nothing much, which was a fine way to spend a Sunday. Now, it wouldn't be long before Mom called them back in the house to enjoy a big dinner, including Bob's favorite thing in the world—Grandma's homemade biscuits. Surprisingly, headlights turned into the drive.

"Who's that?" Bob asked, standing up.

"Oh, crap. It looks like Bonnie's car," growled Tom. Bonnie was their little sister.

"What's she doing home?" Bob wondered.

"Can't be good," muttered Tom.

The evening meal began with the requisite hugs to Bonnie and wrangling the kids to the table. She'd brought her boyfriend along to announce that they were getting married.

"See, that was all, just a new brother-in-law," Bob assured Tom as they were filling plates.

"A no-good-in-law," Tom joked, borrowing a line from a friend. Neither he nor Bob thought Jim was too impressive. Okay—they actually thought he was just plain lazy.

Just as everyone was seated, Dad announced that Bonnie and Jim were moving back and how excited he was to have her home. Then, as Mom wrung her hands, Dad went on, "So, we're going to find Jim a place to work on the farm."

Tom was immediately furious. "Dad! That is ridiculous. We can't …"

Bob knew Tom's temper was headed straight north. "Dad, you know, we just don't have any need right now. Actually, we're sending some of the guys home early for the next month," he explained, trying to diffuse what was going to become a situation.

"Nope, sorry. You boys will find a use for him …"

"That's not gonna happen!" Tom was on his feet, interrupting his dad, toppling a chair. He grabbed the closest thing he could find—to Bob's horror, it was the whole basket of fresh biscuits—and flung it against the wall as he stormed out.

Bob later told me that while he was definitely upset about the fact that he and Tom didn't have a say about bringing Jim on board, he mostly remembers how sad he was to see that whole basket of warm biscuits wasted. That is a pity.

What Does Your Bench Look Like?

As we read in Bob and Tom's story, hiring just any warm bodies, including relatives, is not always the best way to build your bench. It's often perceived as easier to bring family members back and *try* to find a fit for them, rather than going out into the market to recruit new employees. No doubt, it is also difficult to say no to a family member who wants (or *expects*) a job.

Applying this step of the process helps you become more proactive about the roles needed and helps you make more concrete decisions with less emotional involvement. Bob and Tom's family won't be the last family to face this dilemma, but if this has happened in your operation before, why not make it the last time for *your* farm?

Building Your Bench—A Simple Self-Analysis

So, coach, what does the bench look like right now? As you consider it, a good place to start is with three simple categories:

- Age
- Experience
- Expertise

Use a straightforward approach here and list what you have now—a kind of human capital state-of-the-state. Take some time with it, though, and don't overlook any areas of special value or uniqueness. List anything (and anyone!) that is good but could become great with development efforts; save this to review again in Step Four. Of course, include yourself in this analysis.

Age is an obvious question. But, it's not just about the number of years left to work. It's also relevant to consider the specific stage of life a person is in, regardless of age ranges. For example, younger people with small children at home are often pulled in many directions with family commitments. When you are looking to fill positions, are some people more likely to need part-time options or a lot of time off? Do you need to consider two part-time people for one job?

Experience is important when it comes to assessing leadership potential right now and going forward. For example, do your people have the necessary seasons under their belts to take on your role? Again, experience is not just about age. It's also about the type of work a person has done that informs their ability to do a job well. Consider whether people have had the time to navigate a variety of situations, which could help them make informed decisions and avoid the "don't know what you don't know" syndrome.

Your strategy and your personnel development pipeline must align.

Expertise is all about specialty. List the special skills, gifts, talents, or unique training and education of various team members. Again, this is not so much about age as it is about an individual's skills and natural tendencies that the operation can capitalize upon. Analyzing experience can also provide the opposite information, exposing gaps in credentials or training that you should address.

Talent Pipeline

Once you've looked at the current bench, it's time to consider the pipeline needed. I encourage business owners to think broadly about the idea of talent acquisition in terms of a brief risk assessment that considers the talent pool now and the decisions you have yet to make. The light version is below. For my assessment tool that also gives you a readiness score, go to my website at SarahBethAubrey.com

Talent Pipeline Risk Assessment

- I have identified the critical roles for next 12–24 months.

- I am aware of emerging roles in next 24–48 months.

- For emerging roles, I have examined my existing staff for potential growth and development.

- I have a process to manage performance and career paths for key employees.

- I have identified successors for key employees and leaders.

- I have evaluated flight risk and retention strategies to keep key personnel.

It's more critical for smaller sized companies, and especially family businesses, to align strategic direction and future talent needs. An undeveloped talent pipeline poses a greater risk to a company when there are fewer people in line for the CEO role. Your strategy and your personnel development pipeline must align.

Emerging Roles for Innovating Farms

It's imperative to spend quality time thinking strategically about the roles of today, tomorrow, and into the future. Roles at the farm now are different, and the change going forward will blow our minds. There are jobs that don't even exist that will be integral aspects of future farm management. With that in mind, there is no need to develop young leaders into functions that may soon be unnecessary. While you can't avoid essential training to operate the machinery of today or to plant the crop this season, you also need to look ahead and find ways to prepare your successors for the future.

Think of it this way. Training for today's functions is like putting gas in the car to operate it and training and preparing for future capabilities is like designing a whole new vehicle. One is for now, and one is for the future. Both are essential. Yet with the highly reactive nature of farming, even the best operations fall short of doing both. I encourage farm leaders to focus less on needs to be filled season-to-season and more on needs in the 24- to 48-month horizon.

Role Is Different than Job Description

Simply put, the terms job description and role are not the same thing, though they are often use interchangeably. This distinction is both important and useful.

A role is the overall function that a person plays in the business. It may or may not have a descriptive title.

A job description is the list of general and specific activities and responsibilities the person in the role must perform.

Let's use the example of CEO. As the CEO, that's the role you play, and if you call it CEO or president, etc., that is the title. Like an actor in a play, the duties you carry out are the description of how you perform the role. Your unique abilities, the way you "play it," what you put your emphasis on, your natural style and skills—all of these attributes inform the role and what it means in your business.

> I find that the most infrequently used job description is the one for the top leader at the farm. That needs to change as you transition leadership. Have your description ready and clear.

You know how to do your job and what your role is; however, when you consider succession planning, you will need to describe for someone else the duties that are essential. I find that the most infrequently used job description is the one for the top leader at the farm. That needs to change as you transition leadership. Have your description ready and clear.

Decide what roles you need now and what roles you might need in the future, then think about how to slice up and improve the current efficiencies of your employees and leaders. This type of thinking and planning is strategy. Deciding what the job description includes (i.e., the daily tasks and responsibilities for that role) is an executing task. To take a more strategic look at the business's future needs, let's examine roles using the following questions:

- *What functions will be needed to run the business going forward?*

- *What areas of the business are changing, either growing or decreasing?*

- *Are we prepared to meet those changing needs in terms of onboarding new personnel or in terms of time available for current personnel?*

- *Looking at other successful operations, where are they innovating?*

- *What roles today are filled by people who may be transitioning out of the business in five years or less?*

New Roles

New roles are already taking up residence in the most innovative operations. Some of these may already be part of your team.

- Chief Technology Officer
- Chief Financial Officer
- Business Development Director
- Advocacy and Engagement Director
- Safety Director

Sometimes it is easier to think about roles in terms of what you need but don't presently have.

- *What expertise is missing from the business, even if I don't need it today?*
- *How will I meet this need? Is it with training or new hires?*
- *In what ways are we "just getting by"? Are we utilizing people or putting people in positions where they are "just okay"?*
- *Do we have roles now that we don't need or should phase out?*

As a keynote speaker, I am frequently approached at conferences by young leaders sharing some of their transition frustrations. One I often hear is that they know generally *what* they are supposed to be doing at work, but it is hard to see where one job starts and another stops. They are (painfully) aware that this overlap creates conflict—much of it has to do with others in the operation becoming territorial.

Everyone Needs to Know What Everyone Else Does = Role Clarity

To help prevent this, a simple rule should apply—everyone needs to know what everyone else does, aka role clarity. Role clarity should not be reserved for the elite at the top. It's important for every member of the business, down to the most basic laborer. People naturally function better when boundaries are clear.

The younger generation needs to be consulted not just about roles they see as emerging, but also about the training and education they believe is needed. Yes, ask *them*. Do your future a favor and stop dictating now. Seek input from those who will be doing the work of tomorrow.

I Didn't Even Want the Job

In baseball, it would be considered the bottom of the ninth. In politics, it would be called the second reading of the bill before it goes to the governor's desk for signing. In a family business, it's called, well, what can happen in a family business.

It was the day before Thanksgiving, and it was already almost dark as I glanced out of the conference room window. Dreary would have been a *nice* description for the late November day. I was, admittedly, as ready to get home for the holiday as the rest of the people gathered in the local bank's conference room. I thought we were headed in that direction, until during the homestretch of a two-day strategy session with a family farm board and key employees, 35-year-old cousin Justin, finally spoke up. His name had just been added as third-in-line for CEO.

> The younger generation needs to be consulted not just about roles they see as emerging, but also about the training and education they believe is needed. Yes, ask them.

"I don't even want that damn job. Ever. I'm not even sure I want to manage the breeding operation that I do now. I told my wife; I'm thinking about renewing my teaching certificate."

That threw a wrench in the idea of finishing on time ...

How Do You Solve *That?*

You don't, at least not immediately. While we weren't able to come together that day, I agreed to work with Justin one-on-one and work with the rest of the board on an ongoing basis. The first order of business was to figure out what caused the disconnect between what Justin wanted (or didn't) and what the rest of the board had planned for him.

Over time, I learned that Justin had been hired on part-time during the summers to fill in. Though he didn't have a livestock background when he started, he seemed to have a real knack for working with cattle. As the operation expanded, a new heifer development facility was built, and Justin's uncle, Stan, the farm's leader at the time, assumed Justin would rather run the breeding farm than become a teacher and work at school. So, he offered, or, as it turns out, *pushed Justin* into taking the job.

Thinking it would be similar to what he'd been doing over the summers, Justin said yes rather than having to face his Uncle Stan. But he immediately learned that the role came with more expectations than he realized. Stan, a tremendous visionary thinker, hired people and expected Justin to manage them. Stan added a forage enterprise and expected Justin to learn the haying business, though doing so created a conflict with the long-time crop manager who felt *he'd* been slighted by a less qualified family member.

It was never really clear who employees should report to, and Justin, not really enjoying people management, was content to let personnel matters slide and focus on the cows. He was even well-known to play the proverbial ostrich with his head in the sand whenever people got into conflict. They eventually stopped coming to him.

By the time I got to the group, Justin knew that his uncle had a clear succession plan in mind that included Stan's son taking the lead, then Justin's sister, and then him next in line in case of emergency. But he didn't want any part of that. The time he had dreaded for years had come; at the meeting, he had to speak up.

Ultimately, in the case of Justin and his family, what was really missing from the start was role clarity. And as I mentioned above, much of that consternation could have been prevented up front with a clear job description. Maybe knowing what tasks Uncle Stan expected for the role would have forced Justin to have the nerve to speak up sooner. It also may have demonstrated to Uncle Stan what skills were really needed for the job. Maybe he could have divided it up differently and found the right person—someone who did want the responsibility.

Job Descriptions

Job descriptions are simple documents that are a form of agreement between all parties involved. In addition to role clarity, job descriptions can truly help prevent a litany of unpleasantness including, but never limited to, employee confusion, overlap/underlap in tasks, overall inefficiency, arguments at family dinners, hurt feelings, resentment, avoidance of holiday gatherings, and fistfights. Sound like enough reasons? You probably have your own, too.

We all know of examples where there is a lack of role clarity—something like, "Oh, that's Tim. Not sure what he does, but he's been around here for a while." But how about the situation of well-intentioned roles without specific job descriptions? The group of leaders in the following story now acknowledge that their approach started at the very top with solid strategic thinking about roles needed to foster company growth and longevity. Yet what they missed was the exercise of hammering out the tactical job descriptions. As you'll soon read, when both aren't used, problems arise.

What Do You Mean We Can't Deliver the Melons?

The organization was on an exciting path of growth. CEO Jake was diligent about thinking strategically and expanding with purpose. When he started in the business, they were "just veggie farmers" as he put it, doing most of their sales and delivery locally. As he moved into the decision-making role, they added more crops to the lineup and began cleaning, sorting, storing, and categorizing by crops for delivery to wholesale customers. They had added cold storage facilities and a fleet of semis.

Just before I met Jake, another processing facility had been constructed to make canned, ready-to-eat products, and they had built a new local market shop to sell fresh and packaged products to the public. For a while, everyone had just been handling every job coming at them. Jake knew it was chaos, so he hired a consultant and worked with him for six months. They diced up roles into a hierarchy of jobs capable of managing what had become various business units.

Eventually, he created a top-flight C-suite. By making calculated but sweeping changes, he promoted the best people from within and brought in experienced people from the industry. There were now wholesale managers, salespeople for the wholesale and retail units, marketing people to tell the story, production folks for the processing facility, and a new CFO. They promoted good farmers to operations managers and added a human resources department. The plan was well thought out; yet, within the first season, issues emerged faster than lettuce germinating after a spring rain.

So, what was the issue? The roles were clear, but the job descriptions were not. Actually, there were no job descriptions—at all. Jake had deferred to his "top-flight" people to create their own job descriptions. But, of course, they had not written them. No one takes the time to describe their job on paper when it doesn't seem as though anyone else really cares. One of the VPs actually told me that honestly, "No one takes time for that fluff. You just go!" Well, while that may be typical, it also leads to a pretty typical problem symptomatic in growing organizations—chaos.

Suddenly, no one knew what others were doing and weren't even certain what was actually being done! Prior to the new management team, the production guys ran planting and harvest crews and made calls to wholesale buyers. But now, a new wholesale manager had been added. She had big-time grocery experience but no production ag background. As it turned out, she was promising deliveries that the farming side couldn't meet because the weather had not permitted picking. But that was just one example. The list included timing issues between processing and the retail market, and sales reps being told by wholesale accounts that the person they previously dealt with on the farm had already called them—with a different price!

Job roles were clear, but individual responsibilities were not. A lack of job descriptions to clarify the functions to be completed by each person in a given role will lead to overlap and worse.

I once coached an agribusiness CEO, Alan, who was very frustrated with the employees in one key business unit—sales. He constantly said, "They just need to do their jobs!" And he was certainly correct. Interestingly, before being named CEO, he had been the top salesperson at the firm. Not only was he naturally great at sales, but he was also a maverick, the cowboy out there doing what he thought needed to be done.

> **Job descriptions are a simple agreement about what is expected, and many performance issues are fixable with this level of clarity.**

Yet, once he became CEO, sales sagged in his absence. There was a vast difference between what "doing the job" meant to him compared to what it meant to other sales reps. What Alan realized was that he had been seriously pulling the weight for the rest of the sales group. He hadn't noticed the culture that lacked expectations and accountability, and now, it had become painfully clear—sales were down in a great market, and the board of directors was asking their new CEO why.

As we began to work together and he looked at things from a different perch, he realized that while his approach had paid off for him, there was no company-set direction for those with a different style. Under the previous management, there had been no job descriptions. Adding those and building a set of performance reviews to accompany other sales incentives began to turn things around immediately. Some sales reps excelled, and a couple of others realized they'd like to take a different role entirely. But both worked out in the organization's favor.

Job descriptions are a simple agreement about what is expected, and many performance issues are fixable with this level of clarity. There are resources everywhere for building job descriptions, and templates abound. A more detailed checklist for creating a basic job description sheet can be found on my website at SarahBethAubrey.com. For more information on how to properly use job descriptions, head to Step Five.

A Word on Job Titles

An organizational chart with job titles is useful for navigation, even though in the not-so-distant-past, the titles Grandpa, Dad, Mom, and Son *implied* the rank order that everyone understood. Not so now, especially as you bring in employees without an ag background. While you need titles, do you need to be creative about them? Job titles today have started to get really interesting and downright corny. Are they just more accurately descriptive, or are they silly? Do you need a Director of Customer Happiness or a Manure Guru? Maybe not, but titles help people understand the structure, their options for promotion, and the functions of others in the business.

Building Your Bench

Building the bench includes a variety of concerns and factors. A few of the more common are below.

Is Your Operation Attractive?

While the story of Bob, Tom, and Jim may be common, it is just as often a tremendous challenge to attract good employees. Start recruitment with a look around the inside and outside of your business. Be thinking about marketing yourself and your operation. Do you look appealing to prospective hires? Do you offer the best available benefits, work environment, and culture that you can?

Degree or No Degree

What level of education is needed to be a leader in your future business? Is a bachelor's degree or higher necessary for the next CEO? Again, consider what a specific role needs and then look at an individual's capacity to fill it. Perhaps you don't have a college degree, either. So what alternatives are there? Is there a certificate-based program that will enhance critical knowledge from equipment to spreadsheets?

Individuals who are being developed into leaders must be strivers who desire to improve continuously. Even if it is not to gain additional education from a university, an opportunity to be away from the farm and gain exposure to different perspectives and viewpoints allows young leaders to challenge themselves to consider, debate, and adopt new innovations.

Work Away Before Coming Back

A big question on the minds of farmers with young leaders of college age relates to working off farm before coming home. Off-farm opportunities may occur during college or may consist of taking a few seasons after college to work full time somewhere completely different. Does off-farm professional work add value? Should family members be required to work off the farm before coming back full time? I recently moderated a panel with three farmers under age 40, all of whom worked off the farm first. When asked this question, their resounding answer was yes.

One panelist said her science degree and nearly ten years in the medical field were excellent preparation for the work she now does handling soil testing and other agronomic details for the farm. Another producer only spent one year away from the operation after college, but it was in a marketing role. This experience was a huge benefit when the operation expanded, and she found herself handling the zoning, media, and local relations concerning livestock.

The third panelist and his dad had proactively planned for his return and had built a five-year time horizon. However, at year three, Dad found himself without a row-crop manager. After careful consideration, including consulting with their peer group, the dad and son agreed that the timing was right for his return to the farming operation. "It turned out we were really looking for the same thing!" he shared on the panel.

Uh-Oh! They Want to Come Back Before We Planned on It!

Times change. Roles shift, and emergencies happen. And sometimes, young professionals realize that life wasn't greener on the other side of the fence. Even if you've decided to have Junior work off farm before returning, what if the return time horizon changes?

The most important thing for any operation is to be ready to address this issue. Build a family employment policy now. Seek help from your legal professionals and create an approach that is fair and balanced and can be flexible enough to adjust.

Remember the story of Bob, Tom, and Jim? Don't make the mistake their dad made; if you're considering hiring someone in a way that goes against the policy you created, ask yourself why? Is there a truly viable reason beyond the emotional or family pressures you may be facing? Gain input from the emerging leaders in your business. If you respect them, ask them to be part of the decision. It's their future more than it's yours.

Interviewing

Another way to manage and integrate returning family is through a job interview. The panelists I mentioned all interviewed before they came on board. These millennial leaders agreed that a formal interview is important, even for family members—maybe even *especially* for family members. Yes, really.

(Dear Dads over a "certain age," I know there was no such thing as an interview when you joined the farm.

So, what.)

If we acknowledge that the operation is going to look different going forward and will require more professionalism, then the onboarding process must be handled differently than it was a generation ago. Hold the interview in a professional setting, such as the farm office, or even better (since the candidate obviously doesn't need a farm tour!), off-site.

Consider inviting some stakeholders or trusted advisors to participate in the interview process, especially if you don't have multiple layers of other management. They might not make the final decision, but they may be able to provide candid and valuable commentary on blind spots (both positive and negative) that parents can have when seeing their child as an employee.

Assessments in Hiring

More and more companies are using an assessment tool of some kind during the hiring process. Unless the tool is a skill-specific competency test that might be necessary before you can even accept a job application, most firms use assessment tools like CliftonStrengths. Vetting through an application, resume, and possibly a phone interview is standard procedure. But adding assessments to the hiring process is another layer all together.

If you don't presently do it, there are a variety of thoughts about whether using assessments for hiring is a "must-do" or a "nice-to-have." My primary advice is that employers use the same process with new hires that they used with the existing team. For example, if the entire team has used CliftonStrengths and is comfortable with the vernacular of Gallup's 34 Strengths and the Four Domains of Leadership, then it may be appropriate to use this with a new hire to help determine fit. However, it makes no sense to use a tool with someone new if you have nothing to compare it to in the existing business. It's wise to begin using your chosen assessment tool with the existing team first, see how you like it, and then add it to the hiring process.

The Money Game – To Ante Up or Not?

Family members often struggle with how to structure ownership and compensation, a situation that the panelists in my session discussed at length. Everyone had a different take. Some operations believe that family members are employees first (hourly or salary), without a buy-in or own-in opportunity until a specified time. Others created an incentive program tied to performance and provided cash benefits or shares that can be converted to ownership. Still others expect new family members to sign on the dotted line immediately, thus being a part of the "spoils and toils" right from the beginning.

The key to ownership and compensation is consistency. The panelists were clear that no single approach works for everyone. But they also agreed that a structured process that includes input from young farmers fosters the growth they need. Clarity, again, is also essential. People are more motivated and far less frustrated when they understand specifically what they are working toward. The best move? Weigh the options before hiring and build a plan.

Onboarding Process

Do you have a formal onboarding and initial training process in place? If yes, how well is it working? Does it need a refresher? For a short checklist that will help you create and launch a successful onboarding program, head over to my website at SarahBethAubrey.com

Time for the Women in Ag Conversation

Speaking about your bench, as I mentioned in the beginning, it may look a whole lot more feminine than you think.

Why Farming Will Have More Women CEOs

I have a strong belief that the agricultural C-suite at the farm level will look a whole lot different in a generation or less. I fully expect more women not only to end up being farm CEOs but also to garner the top seats at many large agribusinesses as well. Here's why.

Level of Education: In the introduction, I shared the fact that more women graduate from college today than men and that many colleges of agriculture are experiencing a massive shift toward women in both enrollment and graduation for nearly all agricultural degree programs. Visit any college ag campus, and you will probably believe the percentage is much higher. In my home state of Indiana, for example, for the entirety of the 2010s, every State FFA President has been female.

Secondary schools are also doing a better job of encouraging young women to consider the sciences. There is a trend at the college level to bring more young women into STEM-oriented majors and expose them to science-related career options.

Melissa Korn, writing for WSJ.com reports that "women as a share of STEM-degree recipients at the bachelor's level and above increased at nine of the ten largest such programs between 2012 and 2016."[16]

That's a fast advance, especially considering that Korn goes on to write "Six of these programs now award at least one-third of those degrees (STEM-related) to women."

In agriculture, we need educated young people with engineering, coding, programming, science, and math skills. But we will have to fight to get these women to come back to the farm because there are plenty of attractive places for them to work.

Quality of Life and Professional Balance: I speak with a lot of young women who plan to return to the rural area where they grew up, even if it's not to go back to the farm. Reasons range from proximity to family to their enjoyment of the way of life and even the lower cost of living. Yet, most of these women still aim for a career. With essential advances in technology, such as high-speed internet access, rural areas will increasingly have the needed bandwidth to provide more options.

Educated young women today seek a balance of career and family and don't necessarily believe that an hour commute to work each day is worthwhile. Being groomed to lead the farming operation may give them the opportunity to balance family with professional skills and career aspirations.

Fewer (Traditional) Blind Spots: A female expert in grain marketing is fond of saying she prefers to deal with women because they tend to be less close to the "bushel babies" and are less emotional about pulling the trigger on marketing decisions than the guys. Scientific? No. But what she means is that for many men who have traditionally handled the day-to-day planting and harvesting work, it can be challenging to loosen the tie between what the crop looks like and what the market says it will pay. The ability to see the operation from a different background will be helpful to farms of the (near) future.

Advocacy and Relationships: I use the CliftonStrengths assessment when coaching clients and organizations through change. One of the four quadrants of leadership we discuss is about influence—or rather, not just the ability but the desire to influence others. In my experience, this category is typically the one with the fewest strengths—for my male clients, that is.

Again, just my own research, but women in agriculture tend to possess more strengths in areas like advocacy, relationship building, and the need to be inclusive of others than men. Influence is about a variety of things, including the ability to bridge philosophical gaps and negotiation skills. Today that is already important, and going forward, it's mission critical. The ability to negotiate and advocate on behalf of the operation in a way that *non-ag landowners and consumers can relate to* will be a matter of life or death for farms.

What Do You Want in a Successor?

What does the "new you" look like? What does he or she need to know to be successful?

As I asked you in the letter in the beginning, what do you wish someone had taught you?

Considering what you believe is needed in a successor requires taking a candid look at what you have now and what you have in the pipeline. It's about identifying the skills needed that you don't possess and aren't likely to be the one to gain.

Additionally, the opportunity to succeed you someday shouldn't be considered a prize. Consider the perspective of former Citigroup Chairman, Michael O'Neill, who says, "I understand full well that the chairmanship is a trophy for CEOs. I also believe that the non-executive position is probably the right way to govern when you start from scratch. I don't think the chairmanship should be given as a reward or taken away as a penalty."[17]

And, consider the widely speculated notion that GE's handpicked CEO apparently couldn't save the ship. Jack Welch was famous for saying that he gave himself an A, but an F for whom he picked to succeed him.[18]

What letter grade do you want?

What's *Your* Role?

If you are the current leader of the business, what is *your* role? What is *your* job description? As we've discussed, part of the development and growth process for the next generation will require clarity about your role and the expectations that accompany it. If you are the CEO or president, some of the things you do may be obvious. Yet I'll wager that much of it is not, especially to someone who is not experienced enough to recognize critical functions that just seem to happen in your presence.

> **If you are the CEO or president, some of the things you do may be obvious. Yet I'll wager that much of it is not, especially to someone who is not experienced enough to recognize critical functions that just seem to happen in your presence.**

Knowledge transfer to your bench is paramount. Here's an exercise:

Start by writing out the purpose of your role in a sentence or two (or in several bullet points). Then, write out your job description, which should include details about the functions you perform as it stands today. Now, ask yourself the following questions:

- *Will this role be the same for the next person in my position? Why or why not?*

- *How do I see this role needing to change for the next leader?*

- *What are we doing now to develop someone to take on the future leadership role?*

- *What training and experience does the next CEO need to be successful?*

- *What teaching do I need to do now and on a continual basis to prepare the next CEO?*

With your answers in mind, begin or update your overall operational, personnel, and succession strategy to get your people the experience they need and to build the deep expertise required for a successful transition.

This section has been all about the second aspect of Leadership SOI—operational leadership. It targets understanding how to run the business and how to place the right people in the right role.

Operational leadership and building systems now to provide your people with better opportunities for success now and when you turn over the reins. Roles going forward will be different and as unique as every operation that seeks to set themselves apart. Build your bench.

CULTIVATE CULTURE

"Anybody seen Jim?" bellowed Tom as he stormed into the shop. Most of the harvest crew guys just looked at their feet. They knew where Jim was, but they weren't saying.

"Bob! Where is that no-count ...," Tom yelled furiously as he crossed into the office.

"Easy, Tom," Bob stood up from his desk, trying to intervene before Tom blew his lid even more. It was becoming way to common a sight in the last couple months, and frankly, Bob later admitted, he was worried about his older brother.

Tom grabbed a coke from the office fridge and spun around to see Big Ron sitting across from his desk. "What's he doing here?" Tom yelled to Bob. Turning to Big Ron, "What? Now something wrong with the semis? I swear I've got calls to make, and I need to go see the bank about financing for that expansion project. I'm gone half of one afternoon, and that damn Jim ..."

"Jim's not here, Tom, 'cause I sent him home," Big Ron stated flatly. That was a lot of words for Big Ron, the mountain of a man who had worked for Bob and Tom's dad for going on 30 years.

Tom was temporarily stunned into silence, giving Bob a moment to speak. "Tom, sit down a minute. We've got to talk."

Where Coaching and Culture Coexist

Every business has a culture, as does every family. While it's hard to define, culture exists, and it is felt by everyone even though it's not physically seen. What's most interesting about culture is that it is either created or it's simply tolerated. Culture shifts when new leaders take the helm and when new people join the work team. Culture can be changed intentionally with a conscious effort, and it often needs an adjustment when there is damage to repair in relationships or when fissures appear in how a business is run. That's what we see in the story of Tom and Bob and their new brother-in-law, Jim.

The family culture had gotten off kilter when Tom and Bob *mostly* took over the leadership from Dad, yet Dad inserted his will by demanding that they hire Jim. In the habit of their family culture, while they were angry, they didn't *dare* talk to Dad about it. They tolerated Dad's decision—until they couldn't anymore. As you read above, Tom blew up, and Big Ron stepped in to manage the ugly scenario. While the brothers were "in charge," Dad still "ruled."

Instead of trying to talk through a solution, Tom ignored Jim as much as possible, generally angry at him just for being alive. By the time I met the family, Bob and even Big Ron had basically been protecting Jim from Tom and Tom from Jim for months, tiptoeing around the two just trying to keep them from

A culture of coaching and learning is about creating, over time, shared visions. This isn't just consultant talk, it's about sustainability.

seeing each other but not changing anything. They were tolerating the culture of Tom's temper and anger at the situation and Jim's very real incompetence doing the job he had been assigned by Dad. Sadly, the very culture that Tom hated—the way his Dad ruled from a place of anger and demanding silence—was the same way he was acting. The culture they allowed seemed likely to continue for another generation.

Make no mistake. A culture that people believe in, flourish in, and *love* is essential for the long-term health of a farm. A culture of coaching and learning is about creating, over time, shared visions. This isn't just consultant talk; it's about sustainability.

Connectedness

"Creating connection is another form of boundary setting. You are setting a positive boundary, or structure, to form unity. And you are setting a very firm boundary against disconnection and fragmentation," writes Dr. Henry Cloud in *Boundaries for Leaders: Results, Relationships, and Being Ridiculously in Charge*. Dr. Cloud goes on to remind us that leaders can *and must* develop culture. "You get what you create, as well as what you allow. So, create connection and do not allow disconnectedness."[19]

Connectedness is tough to foster, but for leaders, it's just part of the job. Fortunately, process and structure can make things easier and make for good order (think of the adage, "Good fences make good neighbors"). It's imperative to create frameworks that allow individuals the freedom to lead and work in a way that does not stifle their ability or their sense of autonomy. As an executive coach, I have extensively studied coaching strategies, earning professional certifications and even writing a column for the industry magazine *Top Producer*, called "The Farm CEO Coach." But every leader can coach employees successfully.

No, scratch that.

Every leader *must* learn to coach employees successfully.

Being a good coach takes practice and attention. It's challenging to devote hours to working on the business when the workforce is tight, and the margins are slim. Knowing how to operate any enterprise is not automatic, even if you've grown up around it. Challenges today mean young agricultural producers must take on more expansive roles than their farming forbearers. This includes marketing, personnel management, and working with consumers. Successful farms recognize this shift and how these challenges are part of the job. As such, they seek new and unique resources to cultivate young leaders.

When an organization, as many do, revolves around one main decision maker (for many farms that is Dad), intentional coaching allows young leaders to work toward manifesting their own identities and uncovering their unique value. Still, coaching others is not easy. Conversely, it is not easy to receive coaching, either. And it shouldn't be. Coaching should include hard discussions and probing questions that prompt the self-discovery that motivates leaders of any age to be their best and actualize their inner purpose.

Bob, Tom, and even Jim were dealt a situation of having to coexist in a culture that had been allowed and not created. Yet, as you'll soon read, through their efforts, this younger generation worked to emphasize connectedness and created a new culture that set them up for future success.

Train Don't Tell

The old way of telling has now become training. At least, I'm hopeful it has. We are an aging population that cannot replicate ourselves or our knowledge, but we can transfer a great deal of it. We must do it consciously, or it won't happen. Your role is to foster a culture where this transfer of knowledge happens on a regular, focused basis.

Create Your Own Knowledge Transfer Checklist

Sometimes clients ask me where to start coaching. They ask, "What should I coach about?" One way to begin involves creating a knowledge transfer checklist. If you could go back to when you started farming or when you first took over the leadership role, what would you really need to know? What information was lost in the shuffle? Were there things stored away inside Dad's head that you wish you'd known? Create that checklist for your business now.

Generations need to respect each other and constructively learn from each other through intentional, *not occasional,* knowledge transfer. Sure, you've done wills and estate planning, but what about the critical information that needs to be shared? Building a plan for lifelong learning will pay dividends for your future leaders.

Any employer knows that when employees are not engaged, they are likely to leave or, at a minimum, lose productivity. In our business, they could even be unsafe. A Society for Human Resource Management (SHRM) report states that it typically costs the equivalent of six to nine months' wages to find and train a new person. Even for a $12.00/hour hire, that could amount to $20,000 or more.[20]

That's a hit to the bottom line that can be avoided by meeting employees where they are professionally and providing individual coaching when you can.

What *is* Executive Coaching?

To understand how and why to implement a culture of coaching, it makes sense to start with some definitions. Executive coaching is a confidential, one-on-one, professionally facilitated discussion that provides feedback, decision-making support, and an emphasis on accountability. Coaching is used to help the coachee change a behavior, make critical decisions, and to improve performance. Let me tell a story that better defines executive coaching.

> **Executive Coaching is a confidential, one-on-one, professionally facilitated discussion with feedback, decision-making support, and accountability details aimed exclusively at improving performance.**

Accompanied by my brother, Bobby, a lifelong "Cubby," I attended my first MLB game (Major League Baseball, but you already know that). He was excited to share his enthusiasm for the sport. For him, Wrigley Field is a sacred experience. I was just hoping baseball turned out to be more interesting than it looked on TV. Later, while sipping an exorbitantly overpriced cool one, my brother was in his personal heaven on earth. As I detachedly observed the sport, I noticed something that confused me; there was a guy standing by third base wearing the batting team's gear but not running. Risking embarrassment, I queried my brother about it.

"That's the third base coach!" He seemed truly stunned at my ignorance. I reminded him that the last baseball game I had attended was when he and our brothers played little league back when teams were named after the local car dealer, bank, and co-op.

His explanation didn't mean anything to me. Why would a professional ball player need a base coach when his job—running from base to base—seemed so ridiculously obvious?

Holding his hand up to signal that he needed to catch the play before addressing my question, Bobby shook his head. "Ace, (he calls me Ace) every good player needs a base coach. The player *knows* his job is to run home, but he might not be able to see everything, every part of the play. Something might be blocking his view."

I nodded, contemplating the sense in it. "So, it's a strategy then. It helps the player have better vision?"

"Yeah, it's like a new set of eyes. The coach helps him decide when to run hard and when to hold his ground."

> **The coach helps him decide when to run hard and when to hold his ground.**

I wasn't expecting to be enlightened at Wrigley, but as it turns out, baseball is the perfect analogy to describe executive coaching.

Coaching vs. Mentoring

Now, when comparing coaching to mentoring, the main difference is that coaching should be a formal employee-employer kind of work partnership. Or, it's a fee-based arrangement, such as when you bring in a professional.

Mentoring is more typically a personal, long-term relationship between an experienced mentor who has achieved some measure of success with a scenario similar to what the less experienced mentee is going through. There is frequently an age difference, like grandpa and grandson.

Both coaching and mentoring are valuable tools that foster a culture of learning.

10 Differences Between Coaching and Mentoring

Coaching	Mentoring
1. Short-term or for a defined period of time.	1. Often long-term and ongoing.
2. Performance-based and must be measurable.	2. Nurturing approach that, at times, forgives lack of performance.
3. Seeks to enable a concrete, new decision or change, or aims to develop or hone a skill.	3. Provides moral support and not as much training.
4. Must include progress check-ins, often with a report back to a supervisor.	4. Does not require formal reporting or a timely check-in, may have infrequent meetings.
5. A formal process with defined intervals or start/stop.	5. Informal with no process and no determined length of engagement time
6. Provides accountability to guide/develop skills or behavior.	6. Provides a sounding board without defined accountability expectations.
7. Can be fee-based or at least includes a connection to a performance review.	7. Free and doesn't require a penalty or provide a bonus for outcomes.
8. Used in intentional, professional relationships to achieve a desired result.	8. Often based on personal, friendly relationships that provide support.
9. Age of coach or age difference between coach and coachee not relevant.	9. Frequently an age difference exists with the mentor being more seasoned and has experienced the situation of the mentee.
10. Coach is a supervisor seeking performance improvement from employee or a paid professional seeking financial compensation.	10. Mentor is often seeking a mutually gratifying opportunity to give back.

Coaching or mentoring are never a substitute for proper counseling. If you suspect that real mental health support is needed, seek a licensed professional.

Coaching vs. Training – What's the Difference?

One important point of clarification: coaching and training are also different. Training specifically develops a new skill or provides new knowledge or information that allows the employee to perform a function differently. Coaching comes after training. Let's go back to our sports analogy. You first need to be trained on a sport's fundamentals—handling the ball, shooting, swinging, or passing. Once that is understood, then ongoing coaching can hone the craft, continually improving performance over time.

Establishing the Culture of Coaching

Should you outsource all coaching? Not exactly. The best businesses know that success with outside consulting resources happens when there is willingness. It can either be willingness because the business already has a culture of high performance, or it can be willingness to acknowledge a broken culture that needs to change. Either way, there is a willingness to invest time and money to achieve the outcome.

Willingness is also critical for the person being coached. I won't coach a person who doesn't want to be coached. There isn't a perfect way to prescribe a solution to an employee's "problem." Executive coaching is not an intervention; it's an opportunity to learn and develop. If the culture supports growth, personal advancement, learning, and betterment for the individual and for the company, then adding outside coaching programs can be useful when there is a need or a desire.

Fostering a culture of coaching is a leader's role. It's important that employees know that receiving coaching from an internal colleague or an outside resource is not a penalty but rather a privilege. It's essential that employees recognize that the opportunity to be coached means they are valued enough by the organization that it wants to invest in their growth and support their continued development to make them the best team member possible.

> **Executive coaching is not an intervention; it's an opportunity to learn and develop.**

Executive Coaching Opportunity	Not a Good Fit for Executive Coaching
Employee desires professional improvement.	Employee must be penalized or fired.
It's time to discuss a possible career or role realignment.	Employee is behaving irresponsibly or dangerously.
Employee recognizes need for coaching or asks to be coached.	Employee does not recognize need for improvement.
Leader has established with employee the value of coaching as a positive growth opportunity.	Employee is belligerent about being coached, doesn't want to be coached.
Outcomes are expected, clearly stated, and agreed upon.	Last resort before firing someone.

Selecting Key Managers for Coaching:

1. Approach coaching as a positive opportunity for them to grow.
2. Gain commitment and enthusiasm from the individual.
3. Discuss and set specific goals.
4. Set a start and end date.
5. Do a post-assessment evaluation to lock in the accomplishments and determine next steps.

How to Maximize Performance

Executive coaching develops individuals but is also a tremendous way to improve efficiency in teams. The number one benefit of group coaching is trust. Investing in trust building is a solid use of time and money resources. Without trust, performance eventually drops, too.

Group coaching provides a forum for candid but confidential discussions where participants "say it and leave it on the table." It gives the group accountability to each other for improvement in performance or a change in negative behaviors. Group coaching is also a place to build and set communication norms

> **Investing in trust building is a solid use of time and money resources. Without trust, performance eventually drops, too.**

for the group, including response time expectations and other communication practices that may need to be remedied.

Trust Building Pays Off

Besides being angry that they were forced into adding another family member to the farm, ultimately, trust was the baseline issue in the tale of Tom, Bob, and Jim when I met them.

Not long after Big Ron stepped in between the brothers, I met Tom and Bob's wives at a women in ag conference where they invited me out to the farm. By that point, it was more than just biscuits flying. Tom was so mad that Big Ron had stepped in and, without anyone *telling him*, had become Jim's de facto boss, that if he could have thrown Ron off the place, he would have (but, again, Big Ron is a sizeable guy, so he didn't).

After meeting with Bob, Tom, and their wives, the four of them agreed they were seriously ready for a change. Tom didn't want to be angry all the time. He just wasn't sure how to fix the issues that were left to him, so he hadn't. Instead, he had focused his efforts on business building, something he enjoyed and an area where he excelled. He was officially the farm's CEO, and Bob was the CFO. As CEO, all personnel reported up to Tom, even though he actually loathed people management. With nearly 50 employees, he recognized that he didn't spend the time necessary to manage and develop them.

The first thing I did was to interview several employees and family. Even Mom and Dad agreed to be interviewed, though I had to call them in Florida. Then, I met with Jim. I learned that he was deeply unhappy. He'd been given a job running heavy equipment for the group's tiling business but never liked operating machinery. He admitted that he slacked off and missed work days. He didn't like Tom and acted like a jerk around him, but he also acknowledged that it was mostly because Tom didn't like him. He knew that his attitude was poor but felt like it didn't matter since no one wanted him there anyway. The day Big Ron sent him home, he had actually sustained a mild injury that happened, he shared, because he had been browsing social media and wasn't paying attention. Jim was deeply ashamed about being deservedly admonished by Big Ron, the one person he felt treated him decently.

After I'd finished the interviewing process, I met again with Tom and Bob. They strongly desired change yet felt at a loss for not only identifying what to do but how to facilitate the progress they needed. We decided that I would individually coach Tom, Bob, and Jim.

Additionally, as a group, we would have monthly group video conversations. Bob, Tom, and Jim would meet off-site at their accountant's office for the video meetings and, after each call, would then plan a thirty-minute debrief without me. Though they all feared an argument, it didn't happen. I believe this didn't happen for two reasons. First, they truly wanted a fix. Second, they had agreed to come up with a plan for a six month cooling off period where they limited their work time together.

A couple of major items came out of the discussions. First, Tom was happy to have personnel removed from his plate, but Bob was also relieved when I suggested that it not be assigned to him right away, either. He was already very busy managing finances, grain merchandising, and logistics for the trucks and tiling enterprises.

Actually, the company decided to hire Bob's wife, Jenn, as a part-time HR manager to coordinate the employee interface. Though they planned it to be temporary, Jenn was really excited for the project; she had a master's in social work but had taken the last few years away from the workforce to raise children. Once on board, she worked with the brothers and key managers, like Big Ron, to write job descriptions and build a workable organizational chart.

Tom was freed up to focus on business development and felt like a weight had been lifted. His attitude noticeably improved, and he became more pleasant toward Jim almost overnight. As for Jim, even his advocate Big Ron agreed that he should be fired for his lack of performance and poor attitude that bordered on dangerous. And, at the risk of angering Dad, that is what they did.

But the entire group agreed that he might be able to be rehired in a job that they needed filled—part-time mechanic. The company kindly agreed to pay for Jim to attend a two-year vocational program. The first six months required full-time classwork in school. For the next 18 months, he would then work as an apprentice for another company as he finished his training. This transition gave the family some distance as they worked through their personal relationships.

Creating a Group Coaching Forum

A group coaching forum is actually quite simple to organize. Select a group of team members to receive individual professional coaching from an outside resource or perhaps one-on-one mentoring from someone in the organization or industry. Then, coordinate regular group meetings for these team members where a facilitator leads a discussion about progress, challenges, or ideas each team member wants to share and hone with input. As you read above, I started with one-on-one interviews with the individuals.

While this works well if you are using an outside source, it can become personal if you're doing it yourself. In that situation, use a questionnaire to capture issues or concerns. A baseline set of questions can be used for all participants to ensure consistency and can simply ask each person to provide information about goals, concerns, and future plans. From these details, a focus or need is likely to emerge.

It's important that individual responses are not shared with the group! Instead, aggregate the responses to find patterns of agreement and disconnect to use in the group sessions. It is a starting point for managers to develop metrics about what they hope to accomplish with the program. For a group coaching questionnaire, head to my website—SarahBethAubrey.com.

Confidentiality

Confidentiality is key to fostering feelings of trust among your team. Two layers of confidentiality need to be addressed—individual and group. At the individual level, the coach should not discuss one team member's sessions with another. At the group level, each person needs to know that what is discussed during the session is confidential. I usually begin meetings by asking each person to acknowledge, verbally, the confidential nature of the discussion.

Constructive Feedback

Another critical aspect of any healthy culture involves giving and receiving constructive feedback. Younger generations tend to seek feedback more actively than Generations X and older and may feel lost without it. Believe it or not, some people actively seek negative feedback about their work! Feedback about performance is an interesting thing to implement in many family farming cultures where "just do your job" has been an understood fact of life for so long.

However, farming management practices need to adapt and change. Leaders need to recognize that providing constructive feedback can be a valuable growth and learning tool. Leaders who want high performance and decreased costs should consider the merit of a feedback-based approach. When shared, understood, and implemented, feedback can improve performance quickly and reduce errors dramatically. The problem is that, in the past, many leaders provided ineffective feedback. Yelling, the silent treatment, and "told you so" after the fact are certainly *some kind of* feedback, but they don't improve behaviors, change culture, or impact the sustainability of an operation.

Don't "save up" constructive feedback, or it will lose all of its steam. Take the time and apply it immediately when needed. Remember—the purpose of feedback is, whenever possible, to teach and explain. Teaching is demonstrative; it's timely and *shows* how and when something needs to change. Telling is lip service; it *says* the way something should be or worse, should have been done, without context. That is what happens to your feedback message when you hold onto it; the receiver may feel it lacks relevancy and looks more like you're just "riding their case."

When shared, understood, and implemented, feedback can improve performance quickly and reduce errors dramatically.

Best Practices for Using Feedback in Your Culture

1. Don't deliver it in the heat of the moment. If an employee has made a critical error, triage the situation and save the constructive feedback for a time when you are not irate.

2. Deliver the information one-on-one and always include that person's direct manager.

3. If there is a persistent issue, note it in the individual's performance documents and let them know their manager will be circling back at intervals to measure improvement.

4. Ask for buy-in. Use open-ended language such as, "Do you understand?" Also, ask them to describe the desired change in their own words and ask if they have any questions about it. Don't assume they "get it."

5. Discuss and model effective feedback techniques. For the rare employee who loves to get feedback, be sure they don't take an unwarranted license *to give it to everyone else.* To be most effective, the feedback process should be clearly defined and structured.

6. Be specific. Don't speak in negative, broad generalizations like, "You have a poor work ethic." Provide clear examples instead such as, "When I assigned you a recent task, it did not get completed on time, and yet you didn't seem concerned about missing the deadline, what was unclear about the instructions?"

7. Use caution. When providing feedback on delicate topics, be thoughtful in your approach, and have back up. For example, when talking with a female employee about the appropriateness of her attire, you should have at least one other person present for the discussion, especially if you are the opposite gender.

8. It's a gift. Communicate clearly and often that the purpose of feedback is personal development. It is provided because you believe your employees are worth the investment and time. It can be tough to receive feedback, but it's effective when presented in a supportive and actionable way.

Consider the time of day when you give feedback. It is widely acknowledged in HR circles that time of day can have a major impact on whether it's accepted or not. While some might think it better to deliver at the end of the day so you can "send them home to think about it," that may not be the best approach. End-of-day fatigue can naturally lead to bad receptivity. In fact, the best time for a feedback discussion is in the morning when a person's mind is fresh. The day's obligations have not yet started, and stress levels may be lower.

Make a Learning Culture Part of Your Brand

I worked with Bob, Tom, and Jim for 12 months, during which time they realigned a variety of roles, built new systems for accountability and performance, and decided to send several other employees back to school to earn various certifications. This group of young leaders worked extremely hard to completely transform their culture into an operating model that fit their needs and reflected their strengths. About one year after my interactions with them ended, I received a call from Tom.

"Well, I wanted to update you on the funniest thing," he began. "It turns out, we're now back in the business of paying Jim again."

Oh no, I thought. After all the organizing and implementing work we had done together, I'd had so much hope for the family.

"It's not what you might think, which is why I called," he sounded like he was chuckling. (*A good thing...*) "Yeah, we're actually Jim's clients now!"

"I can't believe it," I remarked, maybe a little too honestly.

"Yeah! He worked for us for six months after he finished school, and then one day, he said that he thought he could start his own business running mobile mechanic trucks in our area. At first—can you believe it—I'm like, 'Jim, you can't quit us, you're a damn good mechanic!' But, then, he says, 'You'll be my first client, and I've got another guy I met in school who wants to work.' So, well, we go for it. Now, we're investors in his company, and he's got three trucks."

Wow. Now that's a shift in culture.

Ways to Incorporate Training

- Encourage advanced certifications for operators of specific equipment.
- Enroll your CFO in a grain marketing class.
- Invite industry speakers or sales reps in to give talks during lunch—for everyone, not just the leadership team.
- Attend a one-day industry conference as a group and discuss it on the way home.

A learning culture also pays dividends beyond work culture and productivity. Positive learning habits established in the workplace can improve the presence of your operation in the community by encouraging employees to be excellent stewards of your brand when they're off work. Additionally, CEOs today must constantly differentiate themselves to key stakeholders, landlords, bankers, and consumers.

A learning culture is also a powerful advocacy and public relations strategy. When faced with external criticism and misinformation from the community or press, establishing a unified message at the farm is essential. A focused message is more easily communicated beyond the farm.

Successful operators of tomorrow will endeavor to create a culture where learning and coaching are part of their day-to-day activities— building productivity and sustainability along the way.

BUILD INFLUENCE

Penny was worried about her son, Jared. Her husband Rich insisted she was overreacting. The 26-year-old seemed to have it all together; he had been back working the farm for four years, returning right after college with an agronomy degree. Rich had given him plenty of leeway to manage the employees and even coordinate some of the operations management details, such as work schedules, logistics for seed and chemical deliveries, and equipment maintenance. He'd already scouted many of the fields more accurately and efficiently than Rich ever felt he could have done. Jared had a nice house, a work truck that was paid for as part of his salary, and he was dating a girl in the community.

Still, when Penny finally brought her concerns up during one of their peer group meetings, Rich admitted that he was irked. They had been attending peer network meetings for about three years and were diligent about going. They respected their peer's comments and appreciated giving and receiving input from their group. The day Penny mentioned her "worries" about Jared, Lawrence, one of the older members of their group, looked straight at him and said, "Well, Rich, we've all been wondering when you're gonna let that boy off the place to get some outside exposure."

Rich remembers being stunned into silence as the group all agreed with Lawrence. Maybe Jared needed a change of scenery. Even if he wanted to farm, maybe he needed someone else to talk to besides his parents. They encouraged Rich and Penny to start bringing him along.

Creating a High-Value Peer Network

Many of today's top agricultural producers operate in what the business marketplace calls the "lower middle market," meaning companies with a valuation ranging from $10 million to $50 million. While not Fortune 500, these are competitively sized companies that need to plan for the future. Agriculture is certainly no island out in a sea of corn and wheat. Like similarly sized companies in other industries, we must train and develop people for the future. While corporations, including agribusinesses, spend intentional time exposing emerging leaders to many experiences, at the farm gate, this doesn't happen very often.

Historically, development experiences (like attending events, going to meetings and conferences, serving on boards, and traveling abroad) have been reserved for senior people or saved for a time closer to retirement. That thinking is antiquated. Young leaders need off-the-farm experiences and a more global perspective in order to compete in the current market. It should no longer be considered a privilege to go to a meeting. Rather, it needs to be seen as an essential part of leadership development. External involvement, networking, education, and advising are investments that need to be cultivated and planned. Time away from the farm is time to be maximized.

The Value of Networking

How would you rate the value of your professional network today? If you needed to, could you directly call up the executive director of a statewide industry association? How about the national president of that association? Would your U.S. Senator take your call? If you needed media coverage, do you know the publisher of a major regional newspaper or the producer of your TV station?

What about your successor? Have you introduced her to these people?

More importantly, is she beginning to build her own set of contacts?

While not all of these contacts may seem necessary, knowing who to call when you need support is important. It's called influence. These questions are really asking you to quantify the value of your professional network and are designed to get leaders thinking about how to network with a purpose.

Connections matter for a variety of reasons, including strategic partnerships, lifelong learning, and maintaining awareness of critical market or industry information. If you're one of those types who sees networking as pointless or self-promotional, then you're doing it wrong. It's time to focus on connection.

Value of Professional Connections from Networking

- **Gain new perspectives on the industry.**
- **Discover unique solutions to prevailing problems.**
- **Broaden your outlook.**
- **Share and commiserate about areas of common interest.**
- **Build a resource bank of people with different skills.**
- **Be part of a community.**

How to Network

I understand farm leaders who believe it's more important to stay home, keep their heads down, and work, but all effective leaders need to build their networks to gain influence. After understanding the importance of developing a network, the question I most often hear is, "Okay, so how do I network effectively?"

Board Service as a Networking Opportunity

Serving on a board of directors, either in a paid, elected, or advisory role can provide an excellent and valuable experience for young leaders. As a top producer, you are likely to already hold some kind of board position. So, when the next co-op board or trade association seat comes open, consider stepping aside and nominating your emerging leader for the opportunity. Guaranteed, they need the networking opportunity more than you, and they'll get more mileage out of the opportunity, too. Some of my favorite benefits of board service include:

1. *Enhance your local reputation.*

   Board service changes networking from promotional to purposeful. Your peers get to see you in action, making decisions and critically evaluating situations. It's an opportunity to build trust within the community and to advocate on behalf of agriculture and your farm, especially when you serve on nonagricultural boards.

2. *Gain global travel and second language opportunities.*

   As part of a trade association, you'll have the chance to engage with influential leaders, connect with experts who can help you stay informed, and cultivate relationships that will challenge and encourage you for years.

   Today, many state and national associations bring leaders on trade trips so that they can better understand buyers and to provide a human face to what are sometimes commoditized goods. This is a tremendous chance to travel with purpose and build a better understanding of both agriculture and our global customers.

3. *Develop new skills.*

   Active board members develop critical skills in areas such as leadership and collaboration. They also gain a much better understanding of strategy and planning. As a board member, you may be charged with understanding financials and reviewing balance sheets. These are excellent training opportunities that teach and improve skills that can help you manage your own business entity.

4. *Uncover and change your limiting biases.*

The study of unconscious bias, or the factors and thoughts that inform behavior below our consciousness, is gaining utilization in business circles. We all have these biases and being part of a board allows young leaders to seek out information and norms that are different than how you run your operation. It encourages much bigger thinking.

Conferences, Training, International Travel

While younger readers are high-fiving right now and probably liking my social media pages (*Yes! She's telling Dad to send us on trips!*), I am not talking about vacations. The travel, high-impact events, and conferences I am talking about are meant to provide learning and global experiences. These are more important than ever before. Simply put, global experiences and capabilities are essential if you want to be an agricultural leader into the mid part of this century.

Young leaders need to start developing these global capabilities now. Take an accurate assessment of what you need to improve your global understanding. Do you need language training? Would a cross-cultural communication class help you better understand how to communicate with leaders from other cultures? Search out programs with global industry elements. These

> **NEVER take your professional relationships for granted. Treat them with care and respect because they can have AMAZING powers.**

can be found at leadership academies, university certificate programs, or conferences.

Anyone who aims to be CEO (or one of the top three or four decision makers of their generation) needs global experiences. This is a requirement. Get out in the world and see what our customers see, learn what their needs and preferences are first hand. Find out what they think about us, our products, and our culture. While not everyone wants to run for office or even the state corn board, you need experiences with other leaders where you can engage and learn. Hands down. No excuses.

Travel and Conferences – It's Work not Leisure

Here's how to gain the most during time away from the operation:

- Attend with purpose—what's the game plan?
- Prepare! Learn, read, and/or study in advance.
- Have questions and ideas ready to explore with those you meet.
- Reach out to peers and influencers before you attend and plan meet ups.
- Consciously use the time—be present and focused on what you're learning.
- Find a way to combine travel and additional education. That could mean a trip where you earn a certification, obtain continuing education points, or even work towards a degree.

What is a Peer Advisory Board?

When you need to make a major career decision, to whom do you turn for advice? When you are considering a new project, do you collect unbiased feedback? For industry insights, do you stay informed by talking with smart, savvy people in the know? If you don't have an answer to any or all of these questions, let me provide a possible solution: create your own board of advisors.

A board of advisors (BOA) is simply a small group of professionals whom you respect and trust. The group typically meets in an informal setting (or via phone/video) to help you with career and business decisions. They provide valuable input, feedback, and different perspectives. I've used advisory boards at several intervals throughout my career, and they have, literally, been a game changer every time. An added bonus? An advisory board shows other companies and your employees that your operation has the power to bring together interesting people. That demonstrates influence!

While some professionals call advisory board members "directors," I prefer the term advisor because this position isn't paid, nor does it have any legal right to make official decisions on your behalf. Advisors are more like a group of mentors with relevant experience and a willingness to give solid feedback and to be available if you need them. It's important to note this is not a board of directors with voting or spending power in your organization.

Creating a board of advisors is fairly simple, but you get out of it what you put into it. Here are a few suggestions that I've used and found to work well.

Creating Your Advisory Board

1. *Establish your objectives for creating a BOA.*

   Decide on the purpose for your BOA, then create a short list of objectives. Be as specific as possible, while recognizing that there will be changes along the way. For example, if you are launching a new venture, let your possible advisors know you'll be seeking counsel with start-up questions.

2. *Select BOA members with experience related to your objective.*

   Your purpose and goals should guide your choices for advisors. For instance, if we continue with the business start-up example from above, selecting someone that has launched a company or left a job to start a freelance business would be a good fit. Perhaps someone with a lending or financial background would be helpful. The key tip for selecting BOA members is to avoid friends and family UNLESS they have completely relevant professional experience.

   Avoid friends and family on your BOA unless they have completely relevant professional experience.

3. *Explain the time commitment.*

   The first question anyone is going to ask when you invite them to be a member of an advisory board is about the requisite time commitment. Build the time you believe is necessary into your objectives and be prepared to answer that question up front.

   Also, don't assume that because someone likes you that they will want to be part of a BOA—and don't take offense if they decline your request! Invite people formally, with a call or even better, a detailed email describing what you're looking for that is followed up with a call. You cannot assume that someone you respect would like to serve in this capacity, so you need to ask them in a way that is appealing and professional.

4. *Coordinate regular meetings.*

Unlike a single mentor, a BOA provides you with a customized group of different strengths, experiences, and opinions. As such, I advocate that you facilitate regular connections amongst the members of that group. You could host a quarterly conference call or create an email distribution list. Establish the timing and frequency before asking members to join and provide a brief agenda whenever you gather your BOA.

5. *Give value back.*

I created a BOA for my grant writing company (which was sold to investors in 2015). I know one of the reasons I was able to get such awesome people to participate was that *they also wanted to talk to each other.* During each quarterly meeting or call, we took turns sharing industry insights from our unique arena *before* I asked for help and feedback. The group came to value the connections I offered them through participation as part of my BOA. Once again, it's about influence.

6. *Consider board member term limits.*

People's schedules change, as do your needs. Consider setting board member term limits up front so people can transition on and off the board. A finite time commitment is also a good recruitment tool. I found that I was able to get powerhouse advisors when I had a specific time frame.

For my BOAs, I would ask potential members to join my board for 12 months, and then they could reassess. After the one-year commitment, some stayed, and some moved on. But I found that even those who left offered great suggestions for replacements. A board with set term limits is a great opportunity for younger members in the operation to replace current advisors with key peers from their network. It also helps ensure that your trusted peers and mentors don't feel that their time is being abused.

Peer Group Networks

Back to the story of Penny, Rich, and Jared. Just before the next peer meeting, Jared was surprised when his dad invited him to come along. In the past, it had always been assumed that he would stay behind and keep things going, which was certainly fine. Jared was used to that. But there had also been an underlying current that he hadn't *earned the right* to go on a trip, even if it was for business. Jared wanted to go but was actually a little nervous. He didn't know the other twelve members, though he'd heard a lot about them from his dad. He hoped that he was smart enough to fit in during the discussions.

I met Jared when I was giving a breakout session during a conference aimed at young farmers. I had just asked the group to share any experiences they felt had been particularly beneficial to them. Jared had quickly raised his hand. He told the audience how during his first peer group meeting he had met Chad, a producer from several states away who was about his age.

They had similar backgrounds and plenty to commiserate about since both worked in operations with their dads. Jared said that ever since that first meeting, they've probably talked about once a week on the phone, and they text all the time. They exchange lots of shop talk and agronomy ideas, and Jared wasn't shy about slightly choking up when he shared that the biggest benefit to getting to know Chad and the other guys in the group was that he had stopped feeling so isolated.

Penny had been right; something had been bothering Jared.

I always say … it's not a good peer group meeting unless somebody cries. It has actually become the catchphrase around the table at one of the peer groups I facilitate as part of my company, Elevate Ag (ElevateAg.net). And, just like any inside joke, it stuck because it

> **It's not a good peer group meeting unless somebody cries.**

had impact. Operating a legacy with family at the helm and family as part of the succession plan is most certainly an emotional business. Elevate Ag can help families navigate the management and succession of the operation with our fee-based peer group program. It is made up of farmers from a variety of states with the common goal of learning from each other to improve operationally and build quality relationships.

I create and facilitate peer groups with different structures and forms to meet the needs of individual group members. This means I have facilitated groups that meet three times a year at conferences, groups that only meet at member's operations, and groups that meet solely via video. Sometimes, we have outside speakers, and at other times, we focus heavily on one peer member's individual business challenges. The structure varies, but there are common themes that ensure the peer groups can help members achieve extraordinary results in business and professional development.

I've found that successful peer groups work because they have an agreed upon structure and expectations. There are certain frameworks and tenets that make peer groups work and help hold them together when life gets in the way.

Why Peer Groups Work

1. *Confidentiality*

   Confidentiality is the number one, most important element to agree upon when starting and maintaining a peer group relationship. At a minimum, a formal written policy should be signed by all members. In my groups, we also go around the room before beginning each meeting and verbally acknowledge that what we discuss will remain confidential.

   The reason for such extreme confidentiality is simple; it's about creating an environment where everyone knows they can share openly. It's a place where high achievers and CEOs can combat the "its lonely at the top" feeling and the "I have no one who can relate to me" quandary that surrounds being in a leadership position. The only way to get members to feel comfortable about opening up is to establish a solid foundation of trust.

2. *Feedback*

Each peer group's interaction should include opportunities for members to share what is going on in their businesses and to ask for advice from their fellow members. In my groups, we use operational overviews, which are basically free-form reports that each member gives about the state of the state of their business. Everyone is expected to take a turn at each meeting, and the group is expected to ask challenging questions about that member's progress on items the agreed to work on during the previous meeting. However, there are a variety of ways to encourage sharing during meetings.

3. *Expertise*

One entirely unique aspect of formal peer groups is that the members are the consultants (or experts) in the room. The goal of the group is to learn from each other's expertise and experiences. As a facilitator, sometimes it can be challenging to avoid diving into "consultant mode" (meaning where I am doing the teaching). But the real purpose of the facilitator is to help the members share their relevant experiences and detail practical insights into what has and hasn't worked for them. An effective facilitator guides the conversation and works to encourage members to describe their practiced solutions to problems faced by others in the room. For CEOs, other CEOs are often the very best advisors because they truly understand what's required to lead an operation on a daily basis.

4. *Formal Structure and Outside Facilitation*

You can create your own group, but if you do, I suggest you either create it and become the facilitator or join as a member and find a qualified outside facilitator to run the program. You can't get *what you need* out of a peer group membership if you are concerned with the management of it—booking speakers, moderating the conversation, and planning meal breaks. Let a professional coordinate the details while you reap the rewards of peer advice and counsel.

I also advocate that there be a cost to joining the group. When you commit to something financially, you are much more likely to commit to it professionally as well, which means scheduling time and energy for the group—and yourself. For example, the groups I lead involve fees. Some groups may have an annual fee plus expenses. Others may just charge a fee when booking speakers. There are many options. It's good to note that a fee for involvement differentiates a formal peer group from an advisory board, which does not involve fees.

5. *Accountability*

Successful peer groups have mastered accountability on multiple levels. I've seen groups of executives holding each other accountable for everything from business opportunities and financial improvements to needed changes in a member's health and fitness. As you read in Jared's story, peer groups can even provide the personal connections that improve the mental health of members by offering a secure sense of belonging.

Accountability also means being committed to the group by attending meetings and being present when there. At times, it can be difficult for leaders to step away from work to focus on growth and development, especially when these meetings are states away and involve long drives or flights. Peer groups create a formal structure that supports the idea that time away to work on the business and on the owner's professional growth pays dividends.

6. *Return on Investment*

Countless examples are shared with me each year about how the peer-to-peer experience made or saved members money and returned on their initial investment of time, fee, and travel costs. Quantifying that is as unique as the members themselves, but the value is clear whether it comes from a tip to make a change in the business, the ability to partner with another member on a new business venture, or the chance to avoid mistakes when members share their success (and horror!) stories.

The Power to Give and to Receive

The real secret sauce for successful peer groups involves a willingness to both give and accept valuable, honest counsel. A peer network is not an association or a club. As such, there is not a president or a leader among the members. Maintaining a sense that everyone there is on the same plane is essential, so if you find yourself in a group where it's clear you are the senior member or the standout expert, you need to find a new group where you'll be challenged. If you're not challenged, you're mentoring others and not gaining for *your* business, too. It's also important not to let or enable just one person to dominate the group and emerge as the leader. The word for the group is peer group!

> **A peer network is not an association or a club. There is not a president or a leader.**

A (Few) Words on Appropriate Behavior for Leaders

I've actually considered starting an etiquette class on how to behave as a leader. I should begin by admitting that I'm a stickler for all things polished and professional. I do not wear white shoes (or slacks or carry a white handbag) after Labor Day or before Easter. I know. Many consider this passé, but I consider it civility. My Grandma taught me this rule, and I'm sticking with it.

Appropriateness can be a *difficult* topic in the office because it addresses everything about what is acceptable and what is not. Your job is to steer younger leaders through these waters and help them develop a self-awareness around the topic.

Some of the main areas where appropriate behavior may require coaching include:

- Managing distractions, such as constant selfie-taking or perpetual Apple Watch notifications

- Social media posts, which may be a breach of confidentiality or involve derogatory, inflammatory, or crude content

- Clothing selection. Male employees may dress too casually or in a sloppy manner, while female employees may select outfits that are too casual or revealing

- Careless display of the farm's logo. This may mean wearing dirty logo attire when not at work, parking a logo vehicle somewhere it shouldn't be located, or wearing logo items in public that are torn or unkempt

- General messiness and disorganization, whether that is on the desk, in the truck, or in the equipment

- Romantic relationships with or among employees (perhaps the most sensitive topic of all)

Bottom line, your emerging leaders must be the role models for other employees. Don't excuse them from this responsibility, especially if they are family. The easiest way to help correct issues is to meet them head on—and then have a written policy in place that can be shared with everyone.

Many of these items, especially those related to distractions, are a serious safety matter, but all topics involve a level of respect. In Step One, we discussed establishing a vision and mission. These need to be reflected in all aspects of the operation with all employees representing it to the best of their ability.

Communicating Across Generations

"Siri, do you speak 60-year-old?"

Amy, a dairy farmer from California, told me about the day she returned to her office and literally collapsed at her desk in tears. Things had been bad for a while, but everything had come to a head when her dad yelled at her in front of two employees, saying she needed to have more control over "her people" if she was going to do this job right. Worst of all, one of the guys had been Tony, a long-term employee that Amy felt sure was trying to undermine her at every turn. She had been trying to tell her dad about that concern—and that she needed some kind of help—when he'd blown up. Back at her office and at her wit's end, she literally asked her phone's voice activation assistant, "Siri, do you speak 60-year-old?"

Amy swore the answer came back: "No, I do not understand that."

I actually met Amy through her dad, Roger, who had been a panelist with me at an ag advocacy event. Roger had decided to run for a national board and knew that being elected to serve would take a lot of his time away from the operation for several years. Consequently, he put Amy in charge of a new dairy project and all the personnel who worked at the facility. Besides, the level of automation and the caliber of design in the new facility made his head spin anyway. He said, "Amy could operate that thing from space as long as she had her phone." Roger knew about Amy's struggles, but he was also tremendously proud of her.

"My daughter is excellent, and I gotta' tell you, I am *thrilled* that she came back to the business," he had told me, beaming when he showed me her picture on their social media profile. "She's an engineer, you know. The girl is brilliant."

Roger and I talked a while, and ultimately, he invited me to work with Amy. When I called Amy to confirm her interest, though she had never heard of executive coaching, she was thrilled. Amy related to me that at this point, she was ready to try anything to improve the situation. So, we set about trying to uncover the core issues. First, I did a short questionnaire to get an idea about what was on employees' minds. After gathering this information, I flew out and did seven employee interviews in one afternoon, including interviews with Amy, Roger, and Tony.

What I found out in Tony's interview actually shocked me. Tony had been hired by Roger's grandfather and was actually older than Roger by nearly a decade. Tony was brutally honest, saying that he didn't appreciate being told what to do by someone as young as Amy. He was nearing 70 and was genuinely worried about losing his job, something he couldn't afford to do. He knew his attitude was bad, but he just didn't know how to change it. Besides, Tony shared, sometimes (if he felt like it) Roger would still act like the one in charge, giving orders and moving people around to different tasks. Tony said it was just easier for him to do his own thing and not really listen to either of them!

So, how do you help a young leader manage people who are older than her, who have had many more years of service, and who have never reported to a girl (and don't want to start)? Well, it's not easy, but it starts by looking at the influence and communication. Amy acknowledged that while she "knew her stuff," she lacked the voice at times to stand up for herself and let that be known. She admitted to letting employees that intimidated her do things wrong in the new facility and then sneaking out later to fix the problem when they wouldn't see her do it. Her lack of confidence wasn't about her knowledge. It was about the culture that had fostered a false sense of respect for elders that actually ended up costing her some empowerment and the company efficiency.

Building influence in the industry is important, but it's also essential when you manage people. One of the most critical aspects of influence involves clearly communicating value as well as expectations. While the chain of command may have been unclear in the story of Amy, Roger, and Tony, an additional challenge was that Tony wasn't comfortable communicating his value, either. He had the years of experience and the seasoned knowledge of the industry that Amy did not. Yet, due to his fear about losing his job because he didn't know how to operate the new, higher-tech computer-automated facility, he attempted to protect himself with a poor attitude and by trying to intimidate Amy into leaving him alone. Roger, by default, just let the situation simmer, instead of trying to find the areas of disconnect.

> **If you're a leader in business, realize that all generations need to communicate their value in order to be heard, be effective, and be respected.**

But don't think this is a younger generation problem. It is not. If you're a leader in business, recognize that *all* generations need to communicate their value in order to be heard, be effective, and be respected.

Communicating Value

We have at least three (sometimes four, and soon we might have *five*) generations around our farm workplace! The question is how to get the different generations effectively communicating when there is a span of nearly *a century* between the 14-year-old granddaughter who thinks she'd like to farm and the ninety-something father who still owns much of the land. While part of this involves negotiating skills, it's really about communicating value. What does each person bring to the farm? Recognizing the specific value each member brings to the operation will be essential for this next generation of leaders to understand and manage if they want to succeed.

The baby boomer generation has begun to retire from the workforce, and the Gen Xers have begun or will be moving into the most senior roles. Now, millennials and even Gen Z, those born after 1996, are large parts of the farm workforce, too. Building a work culture that finds common ground among these generations and serves all of them is crucial to creating a workplace that attracts, inspires, and retains top talent. If it isn't already, your farm will need to be such a workplace.

It would probably be much easier if there was just an app for communicating across generations! Sometimes it can feel like we need a translator when we speak to people from other age groups. While time doesn't actually move faster now than it used to when the TV station shut down at night with the playing of the national anthem and started again each morning with the 5 am farm report, it certainly seems as though it does.

There are plenty of resources that cover the perceived differences in the generations, but I won't belabor those here. Rather, let's focus on the solutions. When we work with people across age ranges, we focus on communicating from different frames of reference. Different generations bring different contexts and strengths to the relationship. We need to acknowledge these unique perspectives across generations and work to build lines of communication and understanding. A big part of this involves communicating our unique value to others who may have a different frame of reference.

The Four Big Ways to Communicate Across Generations

A few ways to encourage better communication with others and an appreciation of the value they bring, include:

1. *Acknowledge and appreciate differences (but stop stereotyping!).*

   It can be easy to fall back on generational stereotypes when working across generations, which makes it particularly difficult to correct. However, in these situations, we have to pause and make an effort to avoid these limiting biases. We are all influenced by and, ultimately, products of our time periods—when we were born, raised, became adults, and gained our experiences. These experiences define who we are as individuals and are deeply ingrained in our personalities.

   It's less a matter of changing these intrinsic aspects of ourselves and more about accepting differences and respecting those differences in others. I often counsel groups not to begin by trying to change generational attributes. Instead, I suggest working with individuals to identify specifically the value each person adds to the group. From this foundation, it's a matter of building respect for that person's contribution and role as part of the operation, regardless of their generation.

2. *Understand value differences.*

   "Because it's always been done that way" is the type of thinking that doesn't resonate as a legitimate reason with anyone, and particularly, with people who are new to the business. While different generations may naturally value and prioritize different things, it's valuable to have open discussions and explorations about the reasoning behind why things are done a certain way. It may even lead to everyone agreeing that things might benefit from an update. Effective communication relies on this dialogue and openness to change. Think of it this way. If someone younger questions you about the why of something and you can't immediately explain it, then it is probably worth questioning.

3. *Be willing to learn.*

Younger generations need to be patient and willing to learn from experience. It turns out we actually don't have all the answers just because we turned 21 or graduated from college. In turn, mature generations must have the willingness to teach rather than tell. Do you have an established system for knowledge transfer in your operation? If not, that is what this book is challenging you to do. Pairing a more senior, experienced employee with a junior person is the quintessentially classic way to mentor effectively and give everyone a chance to shine.

4. *Acknowledge differences with respect and expect respect.*

Respect for others needs to be a clear expectation, and managers must set up systems to make this happen. The biggest challenges younger family members have when starting at the farm don't usually involve working with Dad, Grandpa, or even Mom. It's more typical that they will struggle when working with long-term employees who are often like family, but not related.

As we saw with Amy's story, giving millennials and Gen Z employees leadership roles helps them develop critical management skills. But it is a challenge to do this when an older family member isn't quite ready to let go. In these cases, the older family member often undermines the younger member's authority by taking control or changing decisions. Don't do this, even if it's not on purpose.

The best way to avoid this temptation is to create clear reporting structures, which can help prevent a host of avoidable problems. Also, it takes work and vigilance at all levels to ensure that a culture of respect is practiced. Treating others with respect is the best way to establish that culture. Additionally, if you see older employees dissing a young manager, say something. Stepping in under these circumstances sends a message that disrespect will not be tolerated. Respect goes both ways, and it is earned both ways, too.

Cultivate Your Influence

The concept of influence is definitely about industry leadership and taking your own place at the table. This addresses the third aspect of Leadership SOI—industry leadership. In this area, we seek to influence and lead the industry in innovation.

Being an industry leader by developing a high-value peer network and cultivating your influence doesn't have to be something that threatens to become a full-time job. Instead, see it as an opportunity that allows your operation to influence. Perhaps that is the most critical element of all. When we look at global markets and the impact of our local neighbors on how we farm (even *if* we farm in some cases), we must consciously curate influence. We have to be visible, professional, and respected to farm forward from here.

Step Five

DEVELOP ACCOUNTABILITY

Brock was frustrated but afraid to tell anyone about it. He'd been back at the farm for two full years and was beginning his third. While he didn't regret the decision he and his wife, Kristi, had made for him to leave his corporate sales rep job, he was wondering, *just a little*, if they'd made the right choice. Still, feeling frustrated and feeling like admitting that to his dad were two very different things.

"There is always a lot to do, so it's not like I'm bored," he began, quantifying the story he'd just related. We met at a conference where I'd given a presentation. At Kristi's urging, Brock looked me up at the reception.

"What's the root of the frustration?" I asked. "Are you feeling overwhelmed with the business or the workload?"

"No, not that at all. Actually, it's more like *underwhelmed*," he said, sheepishly looking around as if he thought his dad would pop up at any moment and admonish him. "When I was a sales rep, there was always something to shoot for. I mean, you had to hit your goals, or you didn't get bonuses. We had meetings to discuss the next year's rollout. We knew what our objectives were, and I really liked hearing from my boss about what he thought I needed to improve to hit those targets. I knew where I was headed."

"So, are you bored with farming? Is the pace not fast enough for you?"

"No, no way!" Brock insisted. "I came back because I wanted to farm. When Dad approached me, he was so proud of what I'd accomplished and thought I would just know what to do and jump right in. I'm trying hard, and Dad shows me a lot of what I need to know, but we don't ever get into *what* we need to do better or *how* I should be doing. He just says to be patient, and I'll figure it out as he turns stuff over to me. But, honestly, I guess I miss those check-ins, the knowing I was on a certain path towards achieving my next goal. What if I'm not really getting any better?"

What Brock was saying was that he needed something to work towards coupled with directed feedback. Without these elements, he felt adrift. It's actually not uncommon for high-achieving people, call them strivers, to need a measuring stick of some sort, something to work towards. Without expectations, performance inevitably suffers, even from the best of the best.

Developing Top Performers

What's interesting about experience is that the longer we are in business and the more we rise in leadership roles (for example, if we become the CEO of our business), the less we need to know technical skills. Think about it. When was the last time you "turned a wrench"? Maybe you even spend very little time in the tractor or harvester today. No, once you become the farm's primary decision maker, you need business fundamentals. Things like leadership and negotiation skills take precedence over farming know-how.

So, what's the best way to get the leadership skills needed for the top post? Get more training? Head online or back to the university for an MBA? Not bad options if you have the time and budget, but the notion that these kinds of tools are the only way to learn how to be a CEO isn't completely accurate.

Andy Kessler explains, "Despite an industry of business schools, management books and seminars, nobody can teach you how to be an effective boss, leader or entrepreneur. They're all self-taught. I'm always impressed when one is actually successful."[21]

I couldn't agree more. High-potential people with drive, skills, and passion do not a perfect manager make. The most successful people find a way to be effective leaders in their own way, at least in part. That may mean on-the-job experience, reading about business and leadership, taking classes, working with a mentor, or some combination of these options. To be successful, it always starts and ends with you and finding what works for you.

Self-Awareness and Effective Leadership Roles

Effective leaders are remarkably self-aware of their strengths and weaknesses. It's important to know both what we're good at and where we need work so we can leverage our best strengths and play down our lesser virtues. By having this self-awareness, we can potentially predict the aspects of a role that will be easy or difficult for us and whether a specific role is a good fit. Both are important considerations.

Situations arise where we can end up saying yes to a role because we think we should take it or because it seems like the next step like we saw in the story of Justin and his uncle Stan from Step Two. Sometimes in agriculture, there are not enough people to take on all the roles, and it is not only tempting but has been customary for the most senior person of a generation to "be the boss of everybody." That can be a mistake. The question is, how do you begin to realign and solve this situation? One possible approach involves using assessment tools.

Assessments for Self-Awareness

Back in Step Two, I cautioned about using assessments in hiring. Conversely, one of the best ways to place leaders in the right role and to help them gain clarity around issues that might be holding them back is to conduct a 360-degree assessment. Assessments provide a relatively unbiased analysis of an individual, including personality traits, strengths, and weaknesses. And they often provide deeply honest, knee-jerk reactions from the person making the assessment of an employee.

So, what is a 360-degree assessment? It's really just a questionnaire that is designed to be taken by key people in an employee's professional sphere. This should be done anonymously, and while the number of 360's you send out varies by the situation and the person, everyone should answer the same questions about the individual. Questions are designed to uncover specific issues or intents and can provide a baseline competency in certain areas or uncover behavior issues that should be corrected. When reviewing an employee's assessment results with them, take note of any patterns in the responses. These can become a great place to begin a coaching or other professional growth program.

When and how assessments are used needs to be closely linked to their purpose. To start, first decide what you want to accomplish with information gleaned from an employee's peers and supervisors.

Here are some things to consider:

- Is it to help coach someone to improve their self-awareness?

- Is it meant to uncover areas where natural skills could be developed and used to get someone in the right role?

- Will it help build team unity and improve understanding about different work styles or personality types?

Areas of Assessment Caution

- Culling an otherwise good candidate based on an assessment not being in some way "perfect"

- Selecting candidates based almost entirely on how well the assessment fits

- Noticing that the assessment results conflict with what you see and feel about the candidate

- Using assessments sporadically and not across the entire organization

I frequently use 360-assessments with clients and custom design them to meet their needs. For a basic assessment tool example that you could use, head to my website—SarahBethAubrey.com.

There are a couple of important considerations with 360-assessments. First, choose your respondents wisely and always distribute it anonymously. This can be easily accomplished by keying the questions into a free or low-cost online tool like SurveyMonkey and sending participants an email link.

Then, have someone else tally the responses and send them to the individual being assessed so they can view their results. When I deliver 360-assessment results to clients, I encourage them to relax and be open-minded when reviewing the feedback. It is entirely possible that it will not all be positive. I suggest taking some time to sort through the information to look for patterns. It's important to then decide what they want to do with the feedback. You don't have to "listen" to all of it, but you should consider the value in this kind of feedback and how it can make you a better professional.

Performance Metrics

This book is a primer, meant to provide the basics that will get you started developing leaders. After evaluating and assessing where individual employees are today, it's time to turn to expectations, performance, and incentives for the future. That means developing specific job descriptions, as we've already covered. And now, by adding review elements to those descriptions, you can track milestones or lapses in an employee's work. But first, what is performance, and how do you measure it?

Performance = Results.

A very simple equation defines it:

Performance = Results.

What's less simple is what behavior or results are expected to meet the performance desired. That's where the need for clear, defined, and agreed upon expectations comes in. Expectations are different for everyone; they are what high achievers seek and what less internally motivated people need. Having defined targets is like handing employees a map to navigate the job. It's actually a fairly simple process to create and deliver a performance metric.

Clear expectations are what high achievers seek and what less internally motivated people need.

Basics of Building Metrics
- Determine what elements or factors a role needs to accomplish. What specific tasks need to be completed? What are the short- and long-term goals for this role?
- Determine what constitutes below, adequate, exceeds, and exceptional performance for each area. These are the metrics. Write them out in detail.

- Share the metric information with individuals as it relates to their role and with everyone for company-wide performance measures.

- Discuss performance expectations and measurement assessments with each employee to gain buy-in and confirm clarity about performance metrics. It is critical that they understand. Reword any language that is ambiguous or confusing.

- Check in, evaluate, and review metrics with employees at set intervals.

While metrics can clarify what needs to be done and foster buy-in, improved performance relies on explaining why these metrics matter and giving people influence over how they execute tasks. Within reason, allow employees to deliver on performance expectations in their own unique way.

Take the story of ag retail CEO, Alan, whom I mentioned earlier. Alan had come up through the sales team, and as I mentioned before, when he noticed low performing salespeople, he sought a fix. Once he built job descriptions and metrics for his sales group, he noticed improvement. Liking the results he had achieved with the sales team, Alan and the VPs next focused on creating metrics for the branch managers. This group of seven individuals were (he grudgingly acknowledged) a mess of bad attitudes. On top of that, they had a complete unwillingness to help each other. They were the opposite of the "All for one, one for all" attitude he was working to develop.

In the culture under the previous CEO, branch managers had been taught to operate in silos and behave so competitively toward each other that animosity was rampant. Customers knew it, too. When a large customer was threatening to leave the company because he couldn't get what he needed on time, he told Alan: "I can't go to Todd's branch even if Lonnie's branch is out of what I need 'cause if I do, Todd will be so (Bleep-Bleep) ticked off that he'll try to bump me off the applicator's schedule!" Obviously, Alan knew that the branch managers' lack of basic camaraderie had become a customer service problem.

Alan and his operations manager set about cleaning up job descriptions for the branch managers. To foster buy-in and understanding, he decided to invite each one in and worked with them to write job descriptions and metrics for their own employees, including tying those to raises and bonuses, which was something that had never been done before at the company.

Alan started with Henry, the manager of the most profitable branch and actually a friend from high school. The job description piece went over pretty well, but when he started to talk about raises for his branch employees, Henry stunned Alan by saying he had no idea what his employees made. The previous CEO had just told him what to offer (or not) for each employee. When he was asked further about doing a budget, Henry said, "No idea. Never done one. Wasn't allowed."

When Alan shared this with me, he was feeling grim. But as we talked through it during a coaching call, he began to realize that the secrecy that dominated the culture he inherited had contributed to the relationship problems among the branches. They literally never knew what was going to happen to them. They had no influence on the management of their own employees.

Alan moved fast to make changes. He gathered the branch manager group together, and I facilitated a discussion and listening session with only the managers (no Alan or other leadership team members present). Whew! That was a doozy of a day, but Alan's hunch had been right. There was no trust and no team accountability either. He knew he had to change the culture to get performance where it needed to be.

Alan organized a financial training class for the branch managers that incorporated both academic elements from the local university and practical elements taught by a CPA. The classes started from scratch—teaching balance sheets and budgeting principles. He shifted one de facto assistant manager at a lower performing location into a branch-wide logistics manager. This new position worked across all branches to help coordinate product application, oversee equipment and people movement, troubleshoot grower needs, and encourage all branches to pull together when one or another needed extra personnel or inventory.

Finally, Alan tasked me with the creation of a group coaching program with the managers. For four months during the winter season, we met every three weeks. There were some brutal conversations, but little by little, guys started laughing around the table and sharing ideas. One guy, the grumpiest of them all at first, even joked that they were going to miss me when I stopped coming to the conference room and making them "talk about their feelings."

In late spring, Alan was walking across the lot of the home office branch when he saw two chemical applicators drive in from the southernmost location, nearly two hours away. Apparently, they'd been rained out, and their manager dispatched them to come to help out at this location for the next two days. Alan said he literally felt his knees go weak (and was only slightly embarrassed about it). Five months before, he had been ready to fire the whole group. Now, they were working together without him even asking.

Alan and his managers are an example of the connection between role clarity and performance improvement. But, it's also about accountability. Most people want to have accountability for their efforts and success. That was the piece that Alan's managers were missing; all the ability to make decisions and be accountable for them had been taken away, and as such, they'd lost interest in the good of the whole.

Performance improved because people were accountable and bought into their roles.

Bonus and Incentive Structure

Another aspect of performance is about incentivizing people. What do you do to provide motivation for your employees? Do you use incentives and bonuses? If so, how are these working for you? Are you seeing an uptick in performance with the bonuses you give, or are they taken for granted? Do you feel as though your employees have proverbial "tip jars," seeking a bonus *just for showing up?* It may be time to shift from across-the-board bonuses to bonuses only for high performers. Though the traditional "year-end" bonus is still commonly used, the performance-based bonus concept changes the dynamic and rewards people for a job well done, not showing up to a job. If you're thinking of moving to that kind of system, do so with caution and plenty of thought. Here are a few suggestions.

Ideas to Try When Revamping Bonus Structures

Phase It: Going cold turkey on eliminating bonuses is clearly a bad idea for you and employees who may count on the money. When you make a change, consider a phased approach over time and make employees aware in advance.

Scale It: One way to demonstrate transparently who qualifies for a bonus and who doesn't is to create a scale with percentages based on set factors. If an employee exceeds a target, the top tier is met. Similarly, if they miss it entirely, the bonus could be zero.

Make Up for It: Another way to phase in bonus changes it to evaluate performance more than once a year. If at midyear, an employee is not on target for a bonus, let them know and consider providing ways for them to make up for that in the latter half.

Create a New Program for New Hires: It's obviously easier to develop a program from scratch than upsetting the applecart for existing employees. So, start new hires with the structure you would like to use going forward and gradually transition long-term employees into this system.

Merit Isn't Just Sales, Profit, or Efficiency: If tied to the review process, employees could earn a bonus for any set of criteria you create. You could bonus someone (or not) for improving their working relationships with peers, for taking on the responsibility of formally mentoring a new person, or even for putting time into developing the SOPs of their job role. You get to decide.

Developing Effective Leaders & Managers

Decision-Making Tools

You love your equipment, right? New tools are cool! And, just like a shiny JD or Case works perfectly when it's tuned up or brand new, performance notably improves and stays high when employees have functional tools in all areas of the farm. What decision-making tools do you use now? If you're not sure, start by asking employees and key leaders what they could do to be better? Or, ask your next generation what they don't know how to do or what they feel like they would like to learn more about.

As seen in Alan's story, employee input on performance metrics can improve outcomes and buy-in, but what if you have solid job descriptions, role clarity, and metrics *yet still have low performance?* This may be the result of a variety of other issues, but I've found that the most common problem in this scenario involves feeling generally overwhelmed. This overloaded feeling can negatively impact results and morale. One solution is to help people develop better time management skills by providing decision-making tools that can help them effectively prioritize tasks.

Strong performers need to make solid decisions time and time again. In these situations, even ambitious, top-flight people can start to feel overwhelmed when the number of decisions they have to make escalates because they either took on too much or are starting to lose focus and become distracted.

As a CEO, you recognize that you need to focus your efforts and spend time on projects with the largest return on investment. But how do you make that determination? Many times, leaders work all day and feel like they've accomplished nothing. In these situations, leaders need to become more proactive about how they spend their time. This means becoming less reactive to situations by recognizing that not everything is a priority. Using a matrix approach can help guide how to prioritize tasks, based on where they fall in the matrix.

Using this matrix is easy and can fit for any task. You can even draw this out on a napkin. Simply draw one vertical line and one horizontal line across it. Then, put each task item in one of the four boxes that result, with high value-low maintenance at the top left, and low value-high maintenance at the bottom right. You can include nearly any issue or task that you like. For example, you could examine one of your standard operating procedures to determine what aspects need to be shorn up to improve it. Or, use it to redefine certain job roles and then write better job descriptions. There are endless uses, and it's easy for just about anybody.

Below is a simple four-box matrix approach that I use with consulting clients. For more on using matrixes and for a larger template with detailed instructions, visit SarahBethAubrey.com.

Simple Priority Matrix Example

High Value-Low Maintenance	High Value-High Maintenance
Low Value-Low Maintenance	Low Value-High Maintenance

Remember Brock? He was feeling frustrated because he actually wanted performance metrics and feedback on his work. Fortunately, he and his dad communicated pretty well, so when Brock decided to take what he had learned about developing leaders and performance reviews at his corporate gig and implement it at the farm, his dad, Dan, told him to go for it. Brock created metrics for his job and benchmarks for the farm to hit that they *both* were measured on. He even built in some incentives for himself and, believe it or not, for his dad, too! The tangible goals and benefits of achieving them made sense to Brock, and he enjoys working hard to hit the marks.

Not everyone is a striver like Brock, but the bottom line is the bottom line. Ask yourself, what do new hires and developing key decision makers cost you? Have you compared the annual cost of these individuals to their return to margin? Scared to add it up? Remember, hiring and retaining the best, even if the best is family, costs successful operations a lot of money.

So, annuitize your investment and build a structured, results-oriented culture so you can monitor those results. High performers will have a sense of achievement and satisfaction in this defined environment, and lower performers will have clear metrics to shoot for, while you coach their improvement (or invite them to find another role elsewhere). Either way, your operation wins.

Set an Accountability Lead (Hint: It's Probably Not You)

In a busy seasonal and reactive environment, it can be difficult to set up, maintain, and communicate coaching and culture systems. Believe me. I hear it all of the time. Probably the most successful way to ensure these approaches are implemented is by staying connected. Connectivity is the most critical accountability measure. Whether you or your team like it or not, meetings are the best way to stay connected. Since there are always questions about the frequency and content of these meetings, I developed the model that follows.

Example Communication and Connection Model

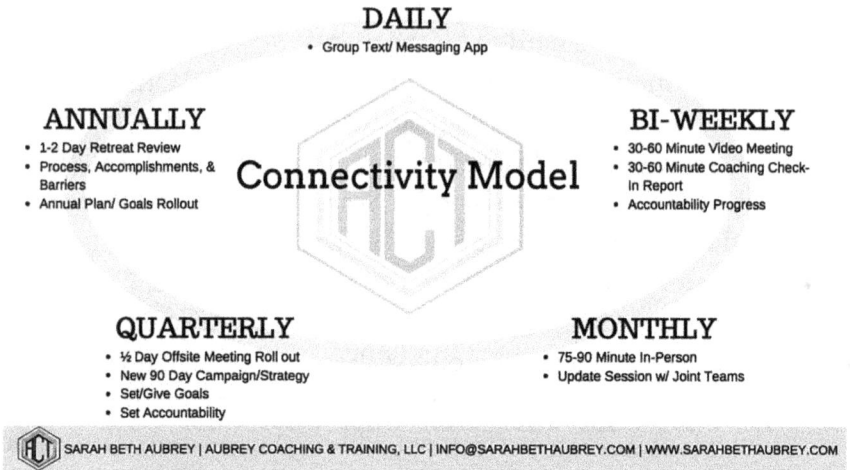

DAILY
- Group Text/ Messaging App

ANNUALLY
- 1-2 Day Retreat Review
- Process, Accomplishments, & Barriers
- Annual Plan/ Goals Rollout

Connectivity Model

BI-WEEKLY
- 30-60 Minute Video Meeting
- 30-60 Minute Coaching Check-In Report
- Accountability Progress

QUARTERLY
- ½ Day Offsite Meeting Roll out
- New 90 Day Campaign/Strategy
- Set/Give Goals
- Set Accountability

MONTHLY
- 75-90 Minute In-Person
- Update Session w/ Joint Teams

SARAH BETH AUBREY | AUBREY COACHING & TRAINING, LLC | INFO@SARAHBETHAUBREY.COM | WWW.SARAHBETHAUBREY.COM

Accountability comes from your team. Self-discipline on the part of the leader is a start, but employees need to feel a sense of responsibility to others. Another way to improve accountability is simply to make it an expectation and create systems to make it a norm. For groups that want a structured format to increase accountability, I have a straightforward worksheet that can be added into performance reviews and job descriptions. This can also be used with a coach, mentor, or separate accountability partner. Accountability partners can be assigned on a short- or long-term basis. This model works great in group meetings as an update or in formal group coaching settings. To take a look at this resource and adapt it to your operation, go to SarahBethAubrey.com.

The Do's and Don'ts of Developing Women Leaders in Your Operation

I've covered a few points about working with emerging young women leaders in agriculture. In this section about accountability and performance, it's a good place do to so again. Take a look:

Do give them P and L experience, but don't call it payroll.

I have heard this very, very good advice several times in recent years from at least three important agribusiness CEOs who were speaking at the front of the room. Two were women, and one was a man, but the advice remained the same. They all shared that to get the top jobs and become successful executing them, all leaders needed some P and L (profit and loss) experience.

Note, they are not talking about the basic accounting work of entering bills and doing the payroll like Mom and Grandma may have traditionally done in the past. That is not P and L, and that does not count. What we're talking about here is true management of the financials, including banking, accounting, and budgeting.

Do let her advocate, but don't think she's just the social media expert.

Just like my mother's generation was told they *could* go to college but should be a teacher, secretary, or nurse, my generation was told we *could* be in agribusiness but that we were best suited for marketing and communications roles. Well, Mom is a teacher, and I'm a writer. How. About. That. Are we products of our generations? Sure, but that doesn't mean that, at least in our cases, we will only do that type of work for the rest of our lives.

Most people don't follow the typical path of starting in one line of work and staying in it for their entire career anymore. That's a notable consideration when developing women in agriculture. For instance, a woman I know from college spent almost fifteen years at a marketing agency. Then, one day, she surprised everyone by leaving the agency to cofound a crop scouting firm. She even went back and finished her agronomy degree at 37. When someone asked her why she made the move, she said that she really wasn't very skilled at marketing campaigns, but when she started school, that was the major that all the girls did.

The current young generation in the workplace is at risk of being placed in a social media box. Don't accidentally box younger millennials and Gen Z females into a social media guru role—and nothing else. Sure, they are naturals at this kind of thing due to their lifelong exposure to social media platforms and the internet, but when they have other talents, make sure your organization benefits from those skills!

Do set up networking opportunities, but don't let her neglect to go.

In a 2015 study of 3,000 women ages 18-64 conducted by professional services firm, KPMG, sixty-seven percent of those interviewed said they needed more professional support to build confidence, but that it was often not available.[22]

You can be the ally who provides your female leaders with designated professional development time away from work to develop the skills they need. That may mean expanding professional networks or building professional confidence.

Remember when we talked about the conferences and travel? Emerging women leaders need to be part of those trips, too. Sometimes, that may take extra encouragement or pushing from top brass. Women frequently find it harder to step away from the mix of family and farm obligations. Make working that out a priority.

Co-Leadership, Different Roles

Some firms, even multi-nationals, use co-CEOs, but I personally don't understand that title. It doesn't seem accurate. If the division of labor is different, and it should be, then the leaders are not co-CEOs. In fact, it may be co-leadership, which I believe is a model that has merit for succession and coaching. Co-leadership is often a solution when two generations are both in charge.

Consider the example from Minnesota of Chuck Sr. and his son, Charlie. During a family planning retreat, the board named Charlie as president, while Chuck Sr. remained CEO. In their succession plan, Charlie becomes CEO when Chuck, Sr. retires, which he's not ready to do yet! Besides, Charlie values having his dad in the same office (except for the two months Sr. now spends in Florida), and he feels supported as he grows into the role of CEO. Charlie is only in his mid-30s, so he has time to learn before he takes over. Additionally, by being president, he already has specific decisions to make with the counsel of his dad and other key business leaders in the family, like his aunt, who serves as a vice president.

This type of leadership structure can only function with open communication and clearly defined roles and responsibilities. In this case, Charlie pursues new acquisitions as well as building and expansion projects. These decisions need to be voted on and approved by the family board before implementation. Both Chuck Sr. and Charlie are voting board members.

Chuck Sr. still oversees all operating divisions and manages relationships with both their large network of landowners and the investors in some of the processing projects. Human resources are handled by the VP. While this process is not perfect, it works because they communicate and are willing to change if needed. In fact, they've revised this structure over the past three years, and the result is a leadership team that is allowing time and expertise to be developed and coached in a very intentional way.

Let Them Lead!
Don't Undermine Your Managers

This is where we return to the story of Amy, her dad, Roger, and the difficult situation with Tony. Several factors created the rift that Amy was so frustrated about. First, although Roger had made Amy the manager of the new dairy unit (and everyone clearly knew it), he didn't give her complete authority. He would still give instructions, and, out of deference and habit, the employees just listened to him. Amy knew that was part of the problem and had tried to explain it to him (though, admittedly, she couldn't speak 60-year-old!). But, also out of deference and habit, Amy didn't want to step in if her dad made or changed rules.

While everyone acknowledged she knew more about the new facility than anyone else, Amy had become afraid to communicate her value and failed to step in and share her opinions, even when her expertise was truly necessary. That led to Tony feeling as though she maybe lacked competence, which couldn't be further from the truth. This was further exacerbated by his own insecurity about being outgunned by a younger person and resulted in his surly attitude.

So, when we met for the second time after the interviews, Roger thought I should be brought on board to do executive coaching with the group. But I wasn't sure we needed to go that far, even though he really liked that idea. Instead, I asked if we could simply have a facilitated, candid discussion with the three of them, and that's when, *through tears from everyone present,* the story above emerged.

Roger agreed to step back and truly let Amy lead. He has been diligent about not undermining her authority, and it has paid off. Even Tony likes her in the role of "boss." He told me during a follow-up video call that she's fair, direct, and easier to work with than Roger ever was! She speaks up when she has an idea and meets with Tony weekly, outside of the larger employee meeting, so they can both stay on track and clear the air if things start to get off course again.

Letting them lead combines all three of the Leadership SOI points. Accountability requires self-leadership and the discipline to take responsibility. Operational leadership is affected by performance at all turns. And with regard to industry leadership, when your farm works to change culture and develop a culture of coaching, mentoring, and high performance, you become a model for others in the industry to follow.

The final key to developing and preparing your emerging leaders to take over means they need to take the reins —at times before they have the seasons under their belts or all of the answers. Let them make the mistakes and, with your support, they'll build on your solid foundations. Ultimately, once you've coached, developed, and provided a solid framework, you've done all you can. Then, it's time to do possibly the most difficult thing of all, step back and watch them lead.

Conclusion

MY WHY AND YOUR CALL TO ACTION

The Farm Sale I Didn't Go To

On a cold, sunny afternoon in mid-March 1997, there was a farm sale in Central Illinois. I knew about it, but I was just, well, sad about the whole thing. It stung, and I was wishing it weren't true. I didn't want to acknowledge that my immediate family's official farming days were over, even though the farmer in question had been working full time at the fertilizer plant for half a decade and trying to do the very difficult part-time, small-time farming thing. I know ... I was being stubborn. But I just didn't want to grow up and believe it, so I didn't go. That farm sale was my Dad's.

Mom and Dad stopped farming as a full-time living at the end of the 1980s. At that point, my Dad kept trying to do it all while keeping up with a busy job selling seed. He continued to farm part time and work full time in the fertilizer industry all while he and my mom had five children, half of us in college. The combination of working in the industry during cropping seasons and trying to get a crop in and out during the same seasons was way too much; it was an exercise in futility and exhaustion.

So, in 1997, they made the smart decision to sell the rest of the equipment and rent the land to neighbors and uncles. But I wasn't there. I didn't go to that sale. I know Dad was disappointed, and I was, too. I assured everyone that I was busy. I was finishing up my senior year at the University of Illinois. But the farm was barely 45 minutes away, and seniors don't have a lot they haven't already finished anyway.

Honestly, it was an identity crisis for me—still is. My dad is a farmer. At least, that is how I always think of him. Even though I didn't want to see it, it was a mistake not to go. I've always regretted it, but I've never regretted working in agriculture, helping farming operations and agribusiness professionals navigate change and build capacity in people. My career so far has been of my choosing, and I choose agriculture and always will. It's my life's work.

This story is likely to be familiar to many of you because you've either experienced it firsthand or know many people who have left full-time farming. Getting out of full-time production agriculture can be a very good decision for a family, as it was for my parents. It's tough, but in the best cases, it's a decision you can make with clarity. However, in many of the stories we hear about, the decision to leave farming is made because no one chose to plan ahead, and things just ended up that way. Worse, it's a decision that is made in an emergency—usually a financial one. That's not what anyone wants, but it is absolutely what can happen without planning.

I wrote this book for those of you who can choose but haven't done so yet. With this book, I'm giving you time back. Call it your five-year warning bell, if you will. Now is the time to look around and make some choices. It's your time. No, actually, that's not quite right. The time you are buying by doing your own version of the exercises and ideas presented in this book goes on the clock of your next generation. Do it for them.

Everyone has a reason or motivation for why their life's work matters and for what they hope to accomplish through their labors. My family's story is why I've always worked in agriculture; why I've built a consulting practice over the last almost 20 years; and why I'm not afraid to push on these issues and say what needs to be said. It's why I lead peer groups and cry with the families just like mine and just like yours. My Why is to make sure that you get the chance to leave your farming life the way you want. I'm asking you to make the time and do this work today, while you have the time to decide how you will leave things for the next generation.

That's My Why.

Your Call to Action

No More Groundhog Day

We all know that feeling; the one where we feel like we're repeating the same day over and over with nothing changing. Do you have a few lingering "groundhog" issues? Perhaps the long-overdue conversation about succession still hasn't been scheduled, or you've been pondering the need to change direction in business but just haven't gotten around to it.

Ultimately, no $20 book is going to be the solution for creating a succession plan for a $20 million operation, but it's my hope that the challenge issued here is one you'll accept. I encourage you to look around right now so that you can seize the opportunity to make the right moves and implement the right culture and practices for the long game. I want to see your operation in 2050.

THANK YOU

Acknowledgments

Special thanks to *Farm Journal* and *Top Producer* magazines for the use of some of my content, which was first published in the Farm CEO Coach column in *Top Producer*. For more on my column, please go to AgWeb.com.

I'd especially like to thank the team at Niche Pressworks for their excellent consulting and strategy work. Most especially of all, I want to thank Nicole Gebhardt for her valuable time.

So many clients and friends have shared their stories—some wishing for them to be included and others wishing for the names to be changed. But every person was generous with their experiences because they believed they could help another farming operation sustain.

Special Note to Readers

I'd like to thank you for the time you've invested in reading this book. If this resource has been of value to you, consider sharing it with others, writing a review on Amazon.com, or sharing one on social media. If you attend industry events that are looking for a speaker, consider sharing this book with the meeting planner and let them know how this topic would be a good program for a conference or national convention.

Be sure to stay connected with me via social media and email. Let me know how this book has helped you, how you've used it, and what tips you have for how other businesses can use it. Let me know your ideas for updates that would be valuable to include in future versions of the book.

If you'd like to learn more about the work I do or are interested in working with my team, connect with me directly to learn about our workshops and one-on-one coaching options.

Lastly, please stay and touch. I would love to work with you directly. You may reach me at info@SarahBethAubrey.com or by connecting with me on social media at the following handles:

Facebook   Aubrey Coaching & Training
Twitter   @SarahBethAubrey
Instagram   @SarahBethAubreysba

Wishing you many happy harvests,
Sarah Beth Aubrey

ABOUT THE AUTHOR

Sarah Beth Aubrey's mission is to enhance the success and profitability in agriculture and rural communities by building capacity in people. She believes everyone has a unique definition of success and strives to foster that potential, whether through one-on-one executive coaching, by facilitating peer groups, or by leading boards through change-based planning initiatives. An entrepreneur since 2003, Sarah Beth founded A.C.T. (Aubrey Coaching & Training, LLC) in 2015, a performance-based leadership training and executive coaching firm. In 2018, she founded Elevate Ag, The Peer Innovation Network, a peer advisory company made up of a network of farmer and agribusiness professional peer groups. She was recently awarded the Indianapolis StartUp Ladies/Women of Skyline Club Transformative Leadership Award for excellence in executive coaching.

Sarah Beth holds a B.S. in Agricultural Communications from the University of Illinois and an M.S. in Strategic Communication from Purdue University. She began her career with Novartis before shifting into business development and training with ABG, Inc. (Adyana).

Entrepreneurial at heart, at 26, Sarah Beth launched a retail meat business, and at 30, she started a grant writing firm. Both businesses have been sold to investors. As the former CEO of Prosperity Ag, LLC, Sarah Beth wrote over 300 successful grants in 39 states, yielding nearly $100 million in funding. She is technically savvy and understands the complex issues facing agriculture and professional development today, blogging on these topics for Forbes.com with the Forbes Coaches Council and writing a column, The Farm CEO Coach, that appears in *Top Producer* magazine.

The author of six books, Sarah Beth is also a member of the Adjunct Faculty at Indiana University-Purdue University Indianapolis (IUPUI) where she designs curriculum and teaches graduate courses on entrepreneurship.

The *Indianapolis Business Journal* named her to 2013's Forty Under 40 list of young professionals to watch, and in 2015, she was honored as a member of Vance Publishing's 40 under 40 in Agriculture. She is a member of Women in AgriBusiness, the Forbes Coaches Council, and serves as a Governor for the board of the Indianapolis-based Skyline Club, a member of the ClubCorp network of professional business clubs.

She is a 2011 alumna of the Richard G. Lugar Excellence in Public Service Series, a member of the National Speaker's Association, the Indiana AgrIInstitute, and the 2015 class of the Stanley K. Lacey Executive Leadership Program. She earned her Certified Executive Coach and Board Certified Coach (BCC) designations in 2019.

As a speaker, Sarah Beth emphasizes personal accountability and process improvement. She works with clients to facilitate professional assessment tools, to create customized small group and individual training needs, and to design qualitative market research programs that support better understanding of customer or member needs and decision-making. Her unique *"Your Strategic Plan on a Page in 1-Hour or Less"* workshop is a popular program for executive teams, farm family governance boards, and conference breakout sessions.

A partial list of clients includes: American Feed Industry Association, Bayer Crop Science, Rabo AgriFinance, Valent, AgriGold, National Association of Farm Managers and Rural Appraiser, BASF, White River Cooperative, the National Sorghum Association, Nutrien, CornPro Trailers, Indiana Farm Bureau, the State of Indiana, Brock Grain Handling, Purdue University, Wabash Valley Power Company, Farm Journal, Vectren, Virginia's Ag Region 2000, NTCA The Rural Broadband Association, and multiple rural cooperative boards.

Sarah Beth and her husband raise cattle in central Indiana where she is active in statewide and local organizations, including serving as director for a rural telecommunications cooperative. Follow Sarah Beth on Facebook, Twitter, LinkedIn, or Instagram. You can also subscribe to her blog at SarahBethAubrey.com.

Endnotes

1   Theresa King and Sue White, *Census of Agriculture Data Now Available,* U.S. Department of Agriculture, April 11, 2019, https://www.usda.gov/media/press-releases/2019/04/11/2017-census-agriculture-data-now-available.

2   Ibid.

3   Ed Maxiner and Sara Wyant, "Big Changes Ahead in Land Ownership and Farm Operators," *Agri-pulse,* February 5, 2019, https://www.agri-pulse.com/articles/11869-big-changes-ahead-in-land-ownership-and-farm-operators.

4   King and White, *Census of Agriculture Data.*

5   Ibid.

6   Rachael Powell, "Ag Careers/Talent-Harvest HR Trends in Agriculture," TalentHarvest (blog), February 12, 2019, https://blog.agcareers.com/talent-harvest/hr-trends-agriculture/.

7   Cheryl Doss and FAO Team, "The Role of Women in Agriculture," ESA Working Paper No. 11-02, March 2011, http://www.fao.org/3/am307e/am307e00.pdf.

8   Powell, "Ag Careers/Talent-Harvest."

9   Colleen Kottkee, "Women Make Up More Than Half of Grads in Ag Programs," *The Wisconsin State Farmer,* May 30, 2018, https://www.wisfarmer.com/story/news/state/2018/05/30/women-make-up-more-than-half-grads-ag-programs/605927002/.

10 Emma Hopkins, "Female Students in Agriculture. Their Increasing Numbers Now the Rule Not the Exception," *Purdue College of Agriculture,* May 12, 2016, https://ag.purdue.edu/agricultures/Pages/Spring2016/02-Female-Students.aspx#.XPJLbtNKjOQ.

11   CoBank, "CoBank Releases 2019 Year Ahead Report–Forces That Will Shape the U.S. Rural Economy," CoBank.com, January 23, 2019, https://www.cobank.com/corporate/news/cobank-2019-year-ahead-report-forces-that-will-shape-the-rural-economy.

12   Coltor Jones and Ruth Simon, "Goodbye George Bailey. Decline of Rural Banks Crimps Small Town Business," WSJ.com, December 25, 2017, https://www.wsj.com/articles/goodbye-george-bailey-decline-of-rural-lending-crimps-small-town-business-1514219515.

13   Rich Horwath, Elevate, *The Three Disciplines of Advanced Strategic Thinking,* (New York: Wiley, 2014).

14   CliftonStrengths, "Understand How Your Talents Work With Others," CliftonStrengths, 2018, https://www.gallupstrengthscenter.com/home/en-us/cliftonstrengths-themes-domains.

15   Greg McKeown, Essentialism. *The Disciplined Pursuit of Less,* (New York: Crown Business, 2014).

16   Melissa Korn.,"As More Women Enter STEM Fields, Difficulties Remain," *Wall Street Journal,* September 25, 2017, https://www.wsj.com/articles/as-more-women-enter-stem-fields-difficulties-remain-1506331800.

17   Telios Demos, "A Turnaround Titan Looks Back," *The Wall Street Journal,* November 23, 2018, https://www.wsj.com/articles/a-turnaround-titan-looks-back-1542978025.

18   Thomas Gryta and Ted Mann, "GE Powered the Century and Then it Burned Out," *The Wall Street Journal,* December 14, 2018, https://www.wsj.com/articles/ge-powered-the-american-centurythen-it-burned-out-11544796010.

19   Horwath, *Elevate.*

20   *Hiring Costs Calculator,* Society of Human Resource Managers, 2019, www.shrm.org.

21   Andy Kessler, "Think you are smarter than your boss?" *Wall Street Journal,* October 14, 2018, https://www.wsj.com/articles/think-youre-smarter-than-your-boss-1539546566.

22 "KPMG Women's Leadership Study: Moving Women Forward into Leadership Roles," KMPG Global, 2015, https://home.kpmg/content/dam/kpmg/ph/pdf/ThoughtLeadershipPublications/KPMGWomensLeadershipStudy.pdf.

www.ingramcontent.com/pod-product-compliance
Lightning Source LLC
Chambersburg PA
CBHW050507210326
41521CB00011B/2364